ADVANCES IN SOME HYPERSONIC VEHICLES TECHNOLOGIES

Edited by **Ramesh K. Agarwal**

Advances in Some Hypersonic Vehicles Technologies
http://dx.doi.org/10.5772/intechopen.68209
Edited by Ramesh K. Agarwal

Contributors

Fang Chen, Hong Liu, Shengtao Zhang, Jian Chen, Zihao Xiong, Zixuan Liang, Chen Bai, Ke Yi, Zhang Ren, Xiaojun Tang, Tianli Hui, Fenglong Yang, Wentao Yu, Juan Yu, Rho Shin Myong, Xiangfa Zhou, Yuriy Shevelev, Giuseppe Pezzella, Antonio Viviani

Published by InTech

Janeza Trdine 9, 51000 Rijeka, Croatia

Publishing Process Manager Dajana Pemac
Technical Editor SPi Global
Cover InTech Design team

First published March, 2018
Printed in Croatia
Legal deposit, Croatia: National and University Library in Zagreb

Additional hard copies can be obtained from orders@intechopen.com

Advances in Some Hypersonic Vehicles Technologies, Edited by Ramesh K. Agarwal
p. cm.
Print ISBN 978-953-51-3903-4
Online ISBN 978-953-51-3904-1

For EU product safety concerns:
IN TECH d.o.o., Prolaz Marije Krucifikse Kozulić 3, 51000 Rijeka, Croatia,
info@intechopen.com or visit our website at intechopen.com.

Meet the editor

Professor Ramesh K. Agarwal is the William Palm professor of Engineering and the director of Aerospace Research and Education Center at the Washington University in St. Louis, USA. From 1994 to 2001, he was the Sam Bloomfield distinguished professor and an executive director of the National Institute for Aviation Research at the Wichita State University in Kansas. From 1978 to 1994, he worked in various scientific and managerial positions at the McDonnell Douglas Research Laboratories in St. Louis. He became the program director and McDonnell Douglas fellow in 1990. He received his PhD degree in Aeronautical Sciences from the Stanford University in 1975, MS degree in Aeronautical Engineering from the University of Minnesota in 1969, and BS degree in Mechanical Engineering from the Indian Institute of Technology, Kharagpur, India, in 1968. He has worked in computational fluid dynamics, computational electromagnetics and acoustics, rarefied gas dynamics and hypersonic flows, flow control, and more recently sustainable air and ground transportation.

Contents

Preface

This book is a compilation of research articles describing the recent advances in some of the hypersonic technologies. The first section of the book consists of four chapters on "Aerodynamic Modeling and Numerical Simulations." The first chapter describes the numerical simulations of rarefied hypersonic flows using the second-order constitutive model of the Boltzmann equation. The second chapter discusses the numerical simulation of hypersonic boundary layer receptivity due to free-stream pulse waves. The third chapter describes the numerical modeling of hypersonic aerodynamics and heat transfer for Martian descent module. The fourth chapter describes an investigative approach at the Phase-A design level for launcher aerodynamics using the CFD and analytical tools. The second section of the book is titled "Miscellaneous Topics." It has three chapters. The first chapter describes the modeling and analysis of fluid-thermal-structure coupling problems for hypersonic vehicles; the second chapter describes the hypersonic vehicles, following LQR design using time-varying weighting matrices; and the third chapter describes the study of porous ceramic phase change composite thermal control systems for hypersonic vehicles. Thus, the book covers a wide variety of topics related to some hypersonic technologies in seven chapters in a single volume. I hope that it can serve as a useful source of reference to both researchers and students interested in learning about certain specific hypersonic technologies.

Ramesh K. Agarwal
Washington University in St. Louis
USA

Section 1

Aerodynamic Modeling and Numerical Simulations

Numerical Simulation of Hypersonic Rarefied Flows Using the Second-Order Constitutive Model of the Boltzmann Equation

Rho Shin Myong

Additional information is available at the end of the chapter

http://dx.doi.org/10.5772/intechopen.70657

Abstract

Various mathematical theories and simulation methods were developed in the past for describing gas flows in nonequilibrium, in particular, hypersonic rarefied regime. They range from the mesoscale models like the Boltzmann equation, the DSMC, and the high-order hydrodynamic equations. The moment equations can be derived by introducing the statistical averages in velocity space and then combining them with the Boltzmann kinetic equation. In this chapter, on the basis of Eu's generalized hydrodynamics and the balanced closure recently developed by Myong, the second-order constitutive model of the Boltzmann equation applicable for numerical simulation of hypersonic rarefied flows is presented. Multi-dimensional computational models of the second-order constitutive equations are also developed based on the concept of decomposition and method of iterations. Finally, some practical applications of the second-order constitutive model to hypersonic rarefied flows like re-entry vehicles with complicated geometry are described.

Keywords: hypersonic rarefied flows, moment equations, balanced closure, numerical simulation, discontinuous Galerkin

1. Introduction

Various mathematical theories and simulation methods were developed in the past for describing gas flows in nonequilibrium, in particular, hypersonic rarefied regime. They range from the mesoscale models like the Boltzmann equation [1–6], the direct simulation Monte Carlo methods [7], and the high order hydrodynamic equations [1–6, 8–20]. Among these models, the kinetic Boltzmann equation plays a central role in the hierarchy of PDE-based mathematical models for gas kinetic theory. The kinetic Boltzmann equation can be transformed into the moment equations by introducing the statistical average in velocity space. Based on the Maxwell's equation of change and the so-called method of moments, Grad [8] in 1949 derived the constitutive equations of viscous shear stresses and heat fluxes from the kinetic Boltzmann equation of the distribution of monatomic gas particles. However, it was found by Grad [9] himself that, within the framework of his constitutive equations, there is a critical Mach number (1.65) beyond which no continuous shock wave solution in high compressive regime is possible.

After Grad's pioneering work in developing gas kinetic theory and subsequent failure of his 13-moment method in describing hypersonic shock structure, there have been enormous efforts to resolve the problem from various perspectives, not only by physicists and mathematicians, but engineers and also chemists. Among such efforts, Eu's works [2–5] to develop the gas kinetic theory consistent with the second law of thermodynamics beyond the linear irreversible thermodynamics stand out. By recognizing the logarithmic form of the nonequilibrium entropy production, Eu [2] in 1980 proposed a canonical distribution function in the exponential form, instead of Grad's polynomial form. He also generalized the equilibrium Gibbs ensemble theory—providing the relationship between thermodynamic variables and the partition functions—to nonequilibrium processes. It turns out that such canonical exponential form assures the nonnegativity of the distribution function and satisfies the second law of thermodynamics in rigorous way, regardless of the level of approximations.

Recently, Myong [15] in 2014 developed a new closure theory which plays a critical role in the development of gas kinetic theory. The new closure was derived from a keen observation of the fact that, when closing open terms in the moment equations derived from the Boltzmann kinetic equation, the number of places to be closed is two (movement and interaction), rather than one (movement only) misled by the Maxwellian molecule assumption in previous theory. Therefore, the order of approximations in handling the two terms—kinematic (movement) and dissipation (interaction) terms—must be the same; for instance, second-order for both terms, leading to the name of the new closure as the *balanced* closure. Then, after applying the Eu's cumulant expansion based on the canonical distribution function to the explicit calculation of the dissipation term and the aforementioned new closure, Myong [15] derived the second-order constitutive models from the Boltzmann kinetic equation and proved that the new models indeed remove the high Mach number shock structure singularity completely, which had remained unsolved for decades.

On the basis of these new theories, this chapter will first describe a recent development in theoretical models for numerical simulation of hypersonic rarefied flows from the viewpoint of the method of moments. It will focus on the detailed derivation of the second-order constitutive model from the original Boltzmann equation and the development of associated computational models for numerical simulation of hypersonic rarefied gas flows in simple geometry as well as complicated real vehicles. Finally, some practical applications of the second-order constitutive model to hypersonic rarefied flows are summarized.

2. The second-order constitutive model of the Boltzmann equation

2.1. The kinetic Boltzmann equation and the method of moments

The Boltzmann equation plays a central role in the hierarchy of mathematical models for gas kinetic theory. It was derived as an evolution equation for the singlet distribution function of a gas by considering the collision dynamics of two particles and combining it with a statistical molecular chaos assumption. Since the molecular chaos assumption is not of a mechanical nature, that is, the Boltzmann equation is based on the assumptions made to "arrive at it" from the reversible Liouville equations of motion, the Boltzmann equation should be regarded as a fundamental kinetic equation at the mesoscopic level of description of macroscopic processes. Thus, it is a postulate for dynamic evolution of singlet distribution functions $f(t,\mathbf{r},\mathbf{v})$ in the phase space (time, position, velocity),

$$\left(\frac{\partial}{\partial t}+\mathbf{v}\cdot\nabla\right)f(\mathbf{v},\mathbf{r},t)=C[f,f_2], \tag{1}$$

which cannot be derived from the pure mechanical deterministic consideration. Although it is a first-order partial differential equation in space and time, its solution becomes very complicated because it is nonlinear owing to the collision integral $C[f,f_2]$, which is made up of products of distribution functions.

The moment equations can be obtained by differentiating the statistical definition of the variable in question with time and later combining with the Boltzmann equation [2–5, 8]; it yields for molecular expressions of general moment $h^{(n)}$

$$\frac{\partial}{\partial t}\left\langle h^{(n)}f\right\rangle+\nabla\cdot\left(\mathbf{u}\left\langle h^{(n)}f\right\rangle+\left\langle \mathbf{c}h^{(n)}f\right\rangle\right)-\left\langle f\frac{d}{dt}h^{(n)}\right\rangle-\left\langle f\mathbf{c}\cdot\nabla h^{(n)}\right\rangle=\Lambda^{(n)}\left(\equiv\left\langle h^{(n)}C[f,f_2]\right\rangle\right). \tag{2}$$

The symbols $\mathbf{c},\mathbf{u},\langle\rangle,\Lambda^{(n)}$ denote the peculiar velocity, the average bulk velocity, the integral in velocity space, and the dissipation (or production) terms, respectively.

2.2. Exact derivation of the conservation laws

The conservation laws of mass, momentum, and total energy can be derived directly from the kinetic Boltzmann equation. For example, in the case of momentum conservation law, differentiating the statistical definition of the momentum with time and combining with the Boltzmann equation yield

$$\frac{\partial}{\partial t}\langle m\mathbf{v}f\rangle=\left\langle m\mathbf{v}\frac{\partial f}{\partial t}\right\rangle=-\langle m(\mathbf{v}\cdot\nabla f)\mathbf{v}\rangle+\langle m\mathbf{v}C[f,f_2]\rangle. \tag{3}$$

Then the first term on the right-hand side becomes

$$-\langle m(\mathbf{v}\cdot\nabla f)\mathbf{v}\rangle=-\nabla\cdot\langle m\mathbf{v}\mathbf{v}f\rangle=-\nabla\cdot\{\rho\mathbf{u}\mathbf{u}+\langle m\mathbf{c}\mathbf{c}f\rangle\}. \tag{4}$$

After the decomposition of the stress $\mathbf{P}$ into the pressure p and the viscous shear stress $\mathbf{\Pi}$ ($[]^{(2)}$ denoting the traceless symmetric part of the tensor),

$$\mathbf{P} \equiv \langle m\mathbf{cc}f\rangle = p\mathbf{I} + \mathbf{\Pi} \text{ where } p \equiv \langle m\mathrm{Tr}(\mathbf{cc})f/3\rangle, \mathbf{\Pi} \equiv \left\langle m[\mathbf{cc}]^{(2)}f\right\rangle, \tag{5}$$

and, using the collisional invariance of the momentum, $\langle m\mathbf{v}C[f,f_2]\rangle = 0$, we obtain

$$\frac{\partial(\rho\mathbf{u})}{\partial t} + \nabla\cdot\left(\rho\mathbf{uu} + p\mathbf{I} + \mathbf{\Pi}\right) = 0, \tag{6}$$

an exact consequence of the original Boltzmann equation. A similar method with the statistical definition of the heat flux, $\mathbf{Q}\equiv\langle mc^2\mathbf{c}f/2\rangle$, can be applied to the derivation of the conservation law of total energy E_t. Then, we obtain the following conservation laws, all of which are an exact consequence of the Boltzmann equation,

$$\rho\frac{d}{dt}\begin{bmatrix}1/\rho\\ \mathbf{u}\\ E_t\end{bmatrix} + \nabla\cdot\begin{bmatrix}-\mathbf{u}\\ p\mathbf{I}\\ p\mathbf{u}\end{bmatrix} + \nabla\cdot\begin{bmatrix}0\\ \mathbf{\Pi}\\ \mathbf{\Pi}\cdot\mathbf{u}+\mathbf{Q}\end{bmatrix} = \begin{bmatrix}0\\0\\0\end{bmatrix}. \tag{7}$$

2.3. Derivation of the second-order and first-order constitutive models via the balanced closure

Starting from molecular expressions of the second-order and third-order moments, the constitutive models of the stress tensor and heat flux vector can be derived via the method of moments and the new balanced closure.

For the second-order moment $h^{(2)}=[m\mathbf{cc}]^{(2)}$, where $m,[]^{(2)}$ denote the mass of gas molecule and the traceless symmetric part, the following constitutive equation of the shear stress tensor $\mathbf{\Pi}\equiv\langle m[\mathbf{cc}]^{(2)}f\rangle$ can be derived from the Maxwell's equation of change (2) [3, 8, 15];

$$\begin{aligned}&\rho\frac{d(\mathbf{\Pi}/\rho)}{dt} + \nabla\cdot\mathbf{\Psi}^{(\Pi)} + 2[\mathbf{\Pi}\cdot\nabla\mathbf{u}]^{(2)} + 2p[\nabla\mathbf{u}]^{(2)} = \mathbf{\Lambda}^{(\Pi)}\left(\equiv\left\langle h^{(2)}C[f,f_2]\right\rangle\right),\\ &\mathbf{\Psi}^{(\Pi)} \equiv \langle m\mathbf{ccc}f\rangle - \langle m\mathrm{Tr}(\mathbf{ccc})f\rangle\mathbf{I}/3.\end{aligned} \tag{8}$$

Similarly, for the next term, $h^{(3)} = (mc^2/2 - mC_pT)\mathbf{c}$, C_p being the heat capacity per mass at constant pressure, the constitutive equation of the heat flux vector $\mathbf{Q}\equiv\langle mc^2\mathbf{c}f/2\rangle$ can be obtained (assuming $\langle m\mathbf{c}f\rangle(\equiv\mathbf{J})=0$ in monatomic gas) [3, 15];

$$\begin{aligned}&\rho\frac{d(\mathbf{Q}/\rho)}{dt} + \nabla\cdot\mathbf{\Psi}^{(Q)} + \langle m\mathbf{ccc}f\rangle\cdot\nabla\mathbf{u} + \frac{d\mathbf{u}}{dt}\cdot\mathbf{\Pi} + \mathbf{Q}\cdot\nabla\mathbf{u} + \mathbf{\Pi}\cdot C_p\nabla T + pC_p\nabla T = \mathbf{\Lambda}^{(Q)}\left(\equiv\left\langle h^{(3)}C[f,f_2]\right\rangle\right),\\ &\mathbf{\Psi}^{(Q)} \equiv \langle mc^2\mathbf{cc}f/2\rangle - C_pT(p\mathbf{I}+\mathbf{\Pi}).\end{aligned} \tag{9}$$

Note that $mC_pT\mathbf{c}$ appears when defining the third-order moments $h^{(3)}$ and both of higher moments $\mathbf{\Psi}^{(\Pi)}$ and $\mathbf{\Psi}^{(Q)}$ vanish near equilibrium. Note also that the constitutive equation (9) was not presented in Grad's original work [8], since his 13-moment (approximate) closure was already applied in the process. In the derivation, the following relations are used;

$$\begin{aligned}
&\left\langle h^{(3)}f\right\rangle = \mathbf{Q} - C_pT\mathbf{J},\ \left\langle \mathbf{c}h^{(3)}f\right\rangle = \langle mc^2\mathbf{cc}f/2\rangle - C_pT\mathbf{P},\ \left\langle f\frac{\partial}{\partial t}h^{(3)}\right\rangle = -\frac{\partial \mathbf{u}}{\partial t}\cdot\mathbf{\Pi} - \mathbf{J}\frac{\partial}{\partial t}(C_pT),\\
&\left\langle f\mathbf{u}\cdot\nabla h^{(3)}\right\rangle = -\rho E\mathbf{u}\cdot\nabla\mathbf{u} - \mathbf{u}\cdot\nabla\mathbf{u}\cdot\mathbf{P} - \mathbf{J}(\mathbf{u}\cdot\nabla C_pT) + \rho C_pT\mathbf{u}\cdot\nabla\mathbf{u},\\
&\left\langle f\mathbf{c}\cdot\nabla h^{(3)}\right\rangle = -\mathbf{Q}\cdot\nabla\mathbf{u} - \Psi^{(P)}\cdot\nabla\mathbf{u} - \mathbf{P}\cdot\nabla C_pT + C_pT\mathbf{J}\cdot\nabla\mathbf{u},\\
&\Psi^{(P)} \equiv \langle m\mathbf{ccc}f\rangle, \rho E \equiv \langle mc^2f/2\rangle, C_pT = p/\rho + E.
\end{aligned} \tag{10}$$

For more details of derivation of heat flux, refer to Appendix A of Myong [15].

Finally, the following constitutive equations, all of which are again an exact consequence of the Boltzmann equation, can be expressed in compact form;

$$\rho\frac{d}{dt}\begin{bmatrix}\mathbf{\Pi}/\rho\\ \mathbf{Q}/\rho\end{bmatrix} + \begin{bmatrix}\nabla\cdot\mathbf{\Psi}^{(\Pi)}\\ \nabla\cdot\mathbf{\Psi}^{(Q)}+\Psi^{(P)}:\nabla\mathbf{u}\end{bmatrix} + \begin{bmatrix}2[\mathbf{\Pi}\cdot\nabla\mathbf{u}]^{(2)}\\ d\mathbf{u}/dt\cdot\mathbf{\Pi}+\mathbf{Q}\cdot\nabla\mathbf{u}+\mathbf{\Pi}\cdot C_p\nabla T\end{bmatrix} + \begin{bmatrix}2p[\nabla\mathbf{u}]^{(2)}\\ C_pp\nabla T\end{bmatrix} = \begin{bmatrix}\mathbf{\Lambda}^{(\Pi)}\\ \mathbf{\Lambda}^{(Q)}\end{bmatrix}. \tag{11}$$

Once the exact constitutive equations are derived, it is necessary to develop a proper closure theory so that they may be applied to the actual calculation of flow problems of practical interests. The closure theory has a long history in many disciplines, since it is essential in describing complex system consisting of vast amount of molecules like fluids, granular media, and soft matter.

Myong [15] recently developed a new theory, so-called balanced closure, by considering the high Mach number shock structure problem. The new closure was derived from a keen observation of the fact that the number of places for closing the exact constitutive equations (11) is two (movement and interaction), rather than one (movement only) misled by the Maxwellian molecule assumption in previous works. In other words, the high order terms associated with molecular interaction, $\mathbf{\Lambda}^{(\Pi)}(\equiv\langle h^{(2)}C[f,f_2]\rangle), \mathbf{\Lambda}^{(Q)}(\equiv\langle h^{(3)}C[f,f_2]\rangle)$, must be taken into account in parallel with the other high order terms arising from movement of molecules $\nabla\cdot\mathbf{\Psi}^{(\Pi)}, \nabla\cdot\mathbf{\Psi}^{(Q)}+\Psi^{(P)}:\nabla\mathbf{u}$. Therefore, the order of approximations in handling the two terms—kinematic (movement) and dissipation (interaction) terms—must be the same; for instance, second-order for both terms.

When this balanced closure is applied to Eq. (11), that is,

$$\begin{bmatrix}\nabla\cdot\mathbf{\Psi}^{(\Pi)}\\ \nabla\cdot\mathbf{\Psi}^{(Q)} + \Psi^{(P)}:\nabla\mathbf{u}\end{bmatrix}_{2nd} = 0,\quad \begin{bmatrix}\mathbf{\Lambda}^{(\Pi)}\\ \mathbf{\Lambda}^{(Q)}\end{bmatrix}_{2nd} = \begin{bmatrix}-p\mathbf{\Pi}/\mu_{NS}\\ -C_pp\mathbf{Q}/k_{NS}\end{bmatrix}q_{2nd}(\kappa_1), \tag{12}$$

the following second-order constitutive equations can be derived;

$$\rho\frac{d}{dt}\begin{bmatrix}\mathbf{\Pi}/\rho\\ \mathbf{Q}/\rho\end{bmatrix} + \begin{bmatrix}2[\mathbf{\Pi}\cdot\nabla\mathbf{u}]^{(2)}\\ d\mathbf{u}/dt\cdot\mathbf{\Pi}+\mathbf{Q}\cdot\nabla\mathbf{u}+\mathbf{\Pi}\cdot C_p\nabla T\end{bmatrix} + \begin{bmatrix}2p[\nabla\mathbf{u}]^{(2)}\\ C_pp\nabla T\end{bmatrix} = \begin{bmatrix}-p\mathbf{\Pi}/\mu_{NS}\\ -C_pp\mathbf{Q}/k_{NS}\end{bmatrix}q_{2nd}(\kappa_1), \tag{13}$$

where the second-order approximation of original dissipation term, $q_{2nd}(\kappa_1)$, can be expressed in a form of hyperbolic sine function whose argument is the first-order cumulant, κ_1, and is given as a Rayleigh dissipation function [3]

$$q_{2nd}(\kappa_1) \equiv \frac{\sinh\kappa_1}{\kappa_1}, \kappa_1 \equiv \frac{(mk_B)^{1/4}}{\sqrt{2}d} \frac{T^{1/4}}{p} \left(\frac{\mathbf{\Pi} : \mathbf{\Pi}}{\mu_{NS}} + \frac{\mathbf{Q} \cdot \mathbf{Q}/T}{k_{NS}} \right)^{1/2}. \tag{14}$$

The symbols k_B, d denote the Boltzmann constant and the diameter of the molecule, respectively. Interestingly, the existence of the hyperbolic sine form in the dissipation (or production) term of second-order constitutive equation can be explained in heuristic way [18, 19] by recognizing that the net change in the number of gas molecules due to the Boltzmann collision integral may be described by gain minus loss, that is, $\exp^{(\text{nonequilibrium})} - \exp^{(-\text{nonequilibrium})}$, so that the leading term of dissipation in the cumulant expansion becomes *sinh*.

Further, it is straightforward to show that, once 1st order approximation (meaning near equilibrium) is introduced to Eq. (13), or equivalently, when the first two terms of the left-hand side are ignored and the right-hand side is taken as first-order ($q_{1st}(\kappa_1)=1$), Eq. (13) recovers the well-known first-order Navier-Stokes-Fourier constitutive equations

$$\begin{bmatrix} 2p[\nabla \mathbf{u}]^{(2)} \\ C_p p \nabla T \end{bmatrix} = \begin{bmatrix} -p\mathbf{\Pi}/\mu_{NS} \\ -C_p p \mathbf{Q}/k_{NS} \end{bmatrix}, \text{equivalently} \begin{bmatrix} \mathbf{\Pi} \\ \mathbf{Q} \end{bmatrix} = \begin{bmatrix} -2\mu_{NS}[\nabla \mathbf{u}]^{(2)} \\ -k_{NS}\nabla T \end{bmatrix}. \tag{15}$$

Lastly, it should be mentioned that, in spite of its conceptual simplicity, "balancing," the new closure theory turned out to be extremely powerful; for example, it can remove the high Mach number shock structure singularity in gas dynamics including hypersonic regime, which had remained unsolved for decades.

2.4. Resolving the high Mach number shock structure singularity

The stationary shock wave structure is a pure one-dimensional compressive gas flow defined as a very thin (order of mean free path) stationary gas flow region between the supersonic upstream and subsonic downstream. The shock wave structure is one of the most-studied problems in gas dynamics, since it is not only important from the technological viewpoint, but it has also been a major stumbling block for theoreticians for a long time after the failure of Grad's 13-moment method in finding continuous shock wave solution beyond a critical Mach number (1.65) [9, 21–23].

The origin of the high Mach number shock structure singularity can be elucidated by investigating the second-order constitutive equations (13) and (14), which are derived based on the balanced closure. Since the mathematical structure of the constitutive equation of heat flux is essentially the same as that of the constitutive equation of shear stress, it is enough to consider the constitutive equation of shear stress only. Then Eq. (13) can be expressed as follows,

$$\rho \frac{d(\mathbf{\Pi}/\rho)}{dt} + 2[\mathbf{\Pi} \cdot \nabla \mathbf{u}]^{(2)} + 2p[\nabla \mathbf{u}]^{(2)} = -\frac{p}{\mu_{NS}} \mathbf{\Pi} q_{2nd}(\kappa_1). \tag{16}$$

When assumptions of one-dimensional flow and steady-state are applied, it can be further simplified into the following algebraic equation [15]

$$-\hat{\Pi}\hat{\Pi}_{NS} - \hat{\Pi}_{NS} = -\hat{\Pi}q_{2nd}\left(\left|\hat{\Pi}\right|\right)\left(= -\hat{\Pi} - \frac{1}{3!}\hat{\Pi}^3 - \frac{1}{5!}\hat{\Pi}^5 - \cdots\right), \text{ where } \hat{\Pi} \equiv \frac{\Pi}{p}, \Pi_{NS} \equiv -\frac{4}{3}\mu_{NS}\frac{\partial u}{\partial x}. \tag{17}$$

This equation shows the nature of the second-order constitutive equation; it provides information of how the stress $\hat{\Pi}$ is determined in the form of $\hat{\Pi}\left(\hat{\Pi}_{NS}\right)$ for a given input $\hat{\Pi}_{NS}$. And it can be easily shown from the solution of the algebraic equation (17) that the equation is indeed well-posed (existence, uniqueness, and continuous dependence on the data) for all inputs, completely free from the shock structure singularity.

On the other hand, when the Maxwellian molecule assumption is introduced in unbalanced way as done in Grad's 1949 work, which is equivalent to assuming $q = 1$ while retaining the

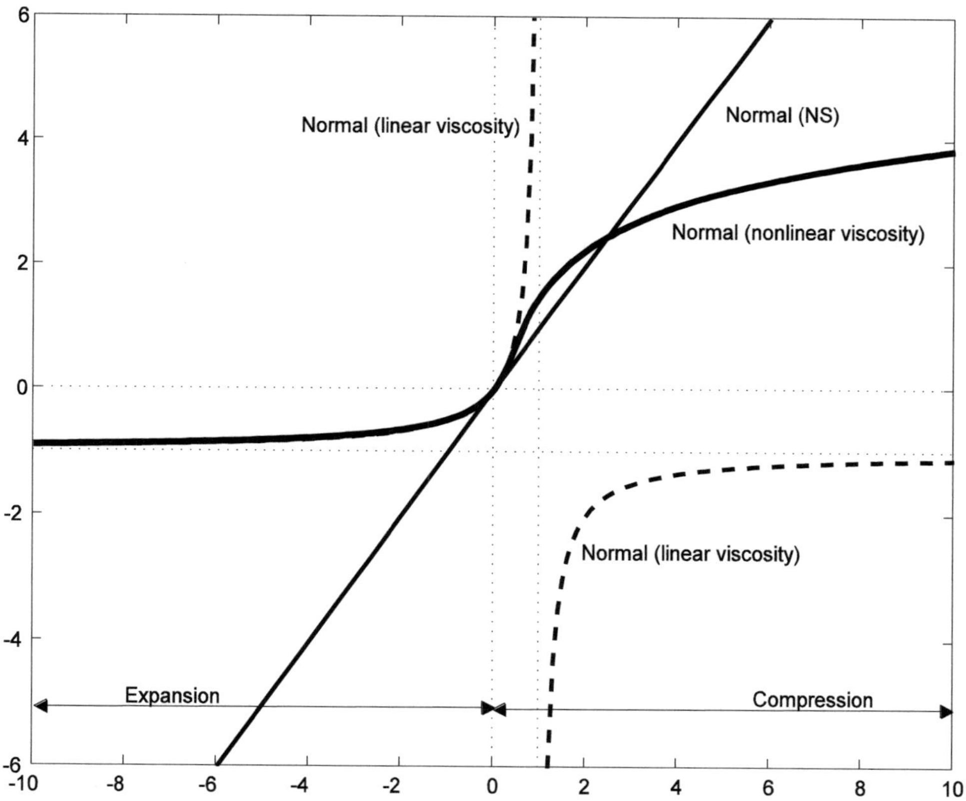

Figure 1. Solutions of the second-order constitutive equation with nonlinear viscosity (17) and the ill-posed constitutive equation (18). The horizontal and vertical axes represent the strain (force) term $\hat{\Pi}_{NS}$ and the normal stress $\hat{\Pi}$, respectively. The gas is expanding in the range of $\hat{\Pi}_{NS} < 0$, whereas the gas is compressed in the range of $\hat{\Pi}_{NS} > 0$. (Reproduced with permission from [15]. Copyright 2014 AIP Publishing LLC).

quadrature term $\left(-\widehat{\Pi}\widehat{\Pi}_{NS}\right)$, then the singularity arises near $\widehat{\Pi}_{NS} = 1$ in the resulting constitutive equation

$$-\widehat{\Pi}\widehat{\Pi}_{NS} - \widehat{\Pi}_{NS} = -\widehat{\Pi}, \text{ or } \widehat{\Pi} = \frac{\widehat{\Pi}_{NS}}{1-\widehat{\Pi}_{NS}}. \tag{18}$$

That is, when the closure (or approximation) is applied in unbalanced way, the high order stress-strain coupling term $|-\widehat{\Pi}\widehat{\Pi}_{NS}|$ of quadratic nature will grow far faster than the thermodynamic force term $|-\widehat{\Pi}_{NS}|$, resulting in an imbalance with the right-hand side term $|-\widehat{\Pi}|$ and eventually a blow-up singularity $\lim_{\widehat{\Pi}_{NS}\to 1} \widehat{\Pi} \to \infty$.

The general solution of the constitutive equation with nonlinear factor q_{2nd} (17) is plotted in **Figure 1** along with the ill-posed equation (18), in the case of Maxwellian molecules. The figure clearly shows that the high order stress-strain coupling term $|-\widehat{\Pi}\widehat{\Pi}_{NS}|$ of quadratic nature plays most important role in the second-order constitutive equation. Interestingly, the figure also shows asymptotic behavior with the increasing degree of expansion, satisfying the free-molecular limit $\widehat{\Pi} \to -1$ or $\Pi + p \to 0$.

3. Numerical simulation of hypersonic rarefied flows

3.1. Computational model for the second-order constitutive model based on decomposition and method of iterations

The second-order constitutive equations (13) are in a form of very complicated partial differential equations so that solving them may be extremely challenging. However, a shortcut is still possible, when we observe that the set of macroscopic variables consists of two subsets, the conserved set and the nonconserved set which vary on two different time scales. It may be estimated that the relaxation times of the nonconserved variables are very short, being of the order of 10^{-10} s. Owing to such a small time scale, on the time scale of variation in the conserved variables, the nonconserved variables have already reached their steady state. Therefore, the constitutive equations (13) of nonconserved variables can be solved with the conserved variables held constant [11–13, 16, 17], and resulting equations are

$$\begin{bmatrix} 2[\mathbf{\Pi}\cdot\nabla\mathbf{u}]^{(2)} \\ -\nabla\cdot(p\mathbf{I}+\mathbf{\Pi})\cdot\mathbf{\Pi}/\rho + \mathbf{Q}\cdot\nabla\mathbf{u} + \mathbf{\Pi}\cdot C_p\nabla T \end{bmatrix} + \begin{bmatrix} 2p[\nabla\mathbf{u}]^{(2)} \\ C_p p\nabla T \end{bmatrix} = \begin{bmatrix} -p\mathbf{\Pi}/\mu_{NS} \\ -C_p p\mathbf{Q}/k_{NS} \end{bmatrix} q_{2nd}(\kappa_1). \tag{19}$$

In general, this algebraic constitutive equations (19) consist of nine equations of $(\Pi_{xx}, \Pi_{xy}, \Pi_{xz}, \Pi_{yy}, \Pi_{yz}, \Pi_{zz}, Q_x, Q_y, Q_z)$ for known 14 parameters $(p, T, \nabla u, \nabla v, \nabla w, \nabla T)$. Because of the highly nonlinear and coupled nature, it is not obvious how to develop a proper numerical method for solving the equations. Nevertheless, it was shown by Myong [11–13] that they can be rather efficiently solved based on the concept of decomposition and method of iterations.

In the case of three-dimensional problems, the stress and heat flux components ($\Pi_{xx}, \Pi_{xy}, \Pi_{xz}$, Q_x) on a line in the physical plane induced by thermodynamic forces of velocity and temperature gradients (u_x, v_x, w_x, T_x) can be approximated as the sum of three solvers: (1) first on (u_x, $0, 0, T_x$), (2) second on ($0, v_x, 0, 0$), and (3) third on $0, 0, w_x, 0$). Hence, nonconserved variables in the case of x-direction can be decomposed as follows;

$$f(u_x, v_x, w_x, T_x) = f_1(u_x, 0, 0, T_x) + f_2(0, v_x, 0, 0) + f_3(0, 0, w_x, 0). \tag{20}$$

Similarly, it is possible to calculate the stress and heat flux in two other primary directions. In the case of y, z-direction, nonconserved variables can be decomposed as follows, respectively,

$$f(u_y, v_y, w_y, T_y) = f_1(0, v_y, 0, T_y) + f_2(u_y, 0, 0, 0) + f_3(0, 0, w_y, 0), \tag{21}$$

$$f(u_z, v_z, w_z, T_z) = f_1(0, 0, w_z, T_y) + f_2(u_z, 0, 0, 0) + f_3(0, v_z, 0, 0). \tag{22}$$

Then, the final value of nonconserved variables ($\Pi_{xx}, \Pi_{xy}, \Pi_{xz}, \Pi_{yy}, \Pi_{yz}, \Pi_{zz}, Q_x, Q_y, Q_z$) can be determined by adding up all these contributions from three decomposed solvers.

Furthermore, it can be noted that three solvers f_1, f_2, f_3 basically consist of two elementary subsets; one on gaseous compression and expansion, and another on the velocity shear flow. Therefore, they can be easily solved by employing the method of iterations, which was first developed by Myong [11–13].

3.2. Explicit modal discontinuous Galerkin (DG) method for high speed gas flows

The second-order algebraic constitutive equations (19), together with the conservation laws (7), are the backbone of the new framework developed for numerical simulation of hypersonic rarefied flows. Because of the nonlinear and coupling nature, a special treatment of viscous terms is required when developing proper numerical schemes. In previous work [16, 17], the mixed DG formulation studied by Bassi and Rebay [24] and other researchers was found suitable for the spatial discretization of the second-order constitutive equations.

The mixed formulation determines the value of the second-order derivatives present in viscous terms by adding auxiliary unknowns **S**, because the second-order derivatives cannot be accommodated directly in a weak formulation using a discontinuous function space. Therefore, **S** can be defined as the derivative of either primitive or conservative variables **U**. This leads to a coupled system of conservation laws for **S** and **U** as

$$\begin{cases} \dfrac{\partial \mathbf{U}}{\partial t} + \nabla \cdot \mathbf{F}_{\text{inv}}(\mathbf{U}) + \nabla \cdot \mathbf{F}_{\text{vis}}(\mathbf{U}, \mathbf{S}) = 0, \\ \mathbf{S}\text{-}\nabla \mathbf{U} = 0. \end{cases} \tag{23}$$

The spatial derivatives of primitive variables can then be computed by expanding the derivatives of the conservative variables; for example,

$$\rho_x = \frac{\partial \rho}{\partial x}, \quad u_x = \frac{1}{\rho}\left[\frac{\partial \rho u}{\partial x} - u\frac{\partial \rho}{\partial x}\right],$$
$$p_x = \gamma(\gamma-1)M^2\left[\frac{\partial \rho E}{\partial x} - \left(u\frac{\partial \rho u}{\partial x} + v\frac{\partial \rho v}{\partial x}\right) + \frac{1}{2}\left(u^2+v^2\right)\frac{\partial \rho}{\partial x}\right]. \tag{24}$$

It was noted by Le et al. [16] that the introduction of an extra set of equations for the auxiliary variables in Eq. (23) is necessary for the nonlinear implicit type of the constitutive models, such as Eq. (19), because it is not possible to directly combine auxiliary equations with primary equations due to the implicitness form of the viscous Jacobian matrix.

In order to discretize the mixed system (23) within the triangulated elements, the exact solutions of $\mathbf{U}$ and $\mathbf{S}$ are approximated by the DG polynomial approximations of $\mathbf{U}_h$ and $\mathbf{S}_h$, respectively,

$$\mathbf{U}_h(\mathbf{x},t) = \sum_{i=0}^{N_k} \widehat{u}_h^i(t)\phi^i(\mathbf{x}), \mathbf{S}_h(\mathbf{x},t) = \sum_{i=0}^{N_k} \widehat{S}_h^i(t)\phi^i(\mathbf{x}), \quad \forall \mathbf{x} \in I, \tag{25}$$

where $\widehat{u}_h^i(t)$ and $\widehat{S}_h^i$ are the local degrees of freedom of $\mathbf{U}$ and $\mathbf{S}$. $\phi(\mathbf{x})$ is the basis function for finite element space, while N_k is the number of required basis function for the k-exact DG approximation. Further, the mixed system (23) is multiplied with the test function, which is taken to be equal to the basis function $\phi(\mathbf{x})$, and then integrated by parts over an element I. This results in the weak formulation of the mixed system for $\mathbf{U}_h$ and $\mathbf{S}_h$

$$\begin{cases} \frac{\partial}{\partial t}\int_I \mathbf{U}_h\phi dV - \int_I \nabla\phi\cdot\mathbf{F}_{\text{inv}}dV + \int_{\partial I} \phi\mathbf{F}_{\text{inv}}\cdot\mathbf{n}d\Gamma - \int_I \nabla\phi\cdot\mathbf{F}_{\text{vis}}dV + \int_{\partial I}\phi\mathbf{F}_{\text{vis}}\cdot\mathbf{n}d\Gamma = 0, \\ \int_I \mathbf{S}_h\phi dV + \int_I \nabla\phi\mathbf{U}_h dV - \int_{\partial I}\phi\mathbf{U}_h\mathbf{n}d\Gamma = 0, \end{cases} \tag{26}$$

where $\mathbf{n}$ is the outward unit normal vector. V and Γ represent the volume and boundary of the element I, respectively. The number of quadrature points necessary for kth order finite element space depends on the type of quadrature rules employed in the numerical process. The Gauss-Legendre quadrature rule has been implemented for both volume and boundary integrations. Therefore, the volume and boundary integrals in Eq. (26) are computed using $2k$ and $2k + 1$ order accurate Gauss quadrature formulas, respectively [25].

The flux functions appearing in Eq. (26) are represented by a numerical flux function. The dimensionless form of the Rusanov (local Lax–Friedrichs (LLF)) flux $\mathbf{h}_{\text{inv}}$ is applied for inviscid terms. This monotone flux is commonly used in the DG method due to its efficiency in computational cost. The Rusanov (LLF) flux is also the most dissipative flux that may improve the stability of DG numerical approximation.

$$\mathbf{F}_{\text{inv}}\cdot\mathbf{n} \approx \mathbf{h}_{\text{inv}}\left(\mathbf{U}^-,\mathbf{U}^+\right) = \frac{1}{2}\left[\mathbf{F}_{\text{inv}}\left(\mathbf{U}^-\right) + \mathbf{F}_{\text{inv}}\left(\mathbf{U}^+\right) - C\left(\mathbf{U}^+ - \mathbf{U}^-\right)\right],$$
$$C = \max\left(|\mathbf{u}^-| + a_S^-, |\mathbf{u}^+| + a_S^+\right). \tag{27}$$

Here a_S is the speed of sound at an elemental interface, and the superscripts (+) and (−) denote the inside and outside sides at an elemental interface. The central flux (BR1) [24] is employed as the numerical fluxes for calculation of auxiliary and viscous fluxes at elemental interfaces;

$$\begin{aligned} \mathbf{F}_{\text{vis}} \cdot \mathbf{n} \approx \mathbf{h}_{\text{vis}}\left(\mathbf{U}^-, \mathrm{S}^-, \mathbf{U}^+, \mathrm{S}^+\right) &= \frac{1}{2}\left[\mathbf{F}_{\text{vis}}\left(\mathbf{U}^-, \mathrm{S}^-\right) + \mathbf{F}_{\text{vis}}\left(\mathbf{U}^+, \mathrm{S}^+\right)\right], \\ \mathbf{U} \cdot \mathbf{n} \approx \mathbf{h}_{\text{aux}}\left(\mathbf{U}^-, \mathbf{U}^+, \mathbf{n}\right) &= \frac{1}{2}\left[\mathbf{U}^- + \mathbf{U}^+\right]. \end{aligned} \tag{28}$$

By assembling all the elemental contributions together, the semi-discrete DG formulation for conservation laws (23) yields a system of ordinary differential equations in time for each element as

$$\mathbf{M}\frac{d\mathbf{U}}{dt} = \mathbf{R}(\mathbf{U}), \tag{29}$$

where $\mathbf{M}$ is the diagonal mass matrix and $\mathbf{R}(\mathbf{U})$ is the residual vector of the system. A third-order total variation diminishing Runge-Kutta (TVD-RK) method is employed for explicit time marching. The local time step for each element is determined by the following relation

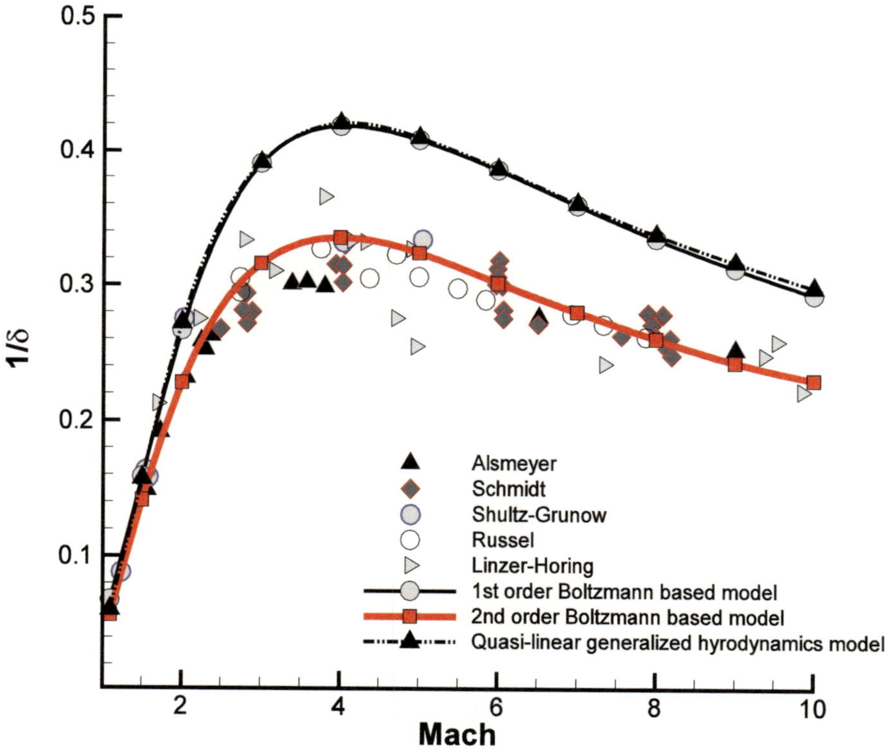

Figure 2. Comparison of the inverse shock density thickness of argon gas.

$$\Delta t = \frac{h}{(2k+1)} \frac{CFL}{|u| + a_s + \frac{1}{\mathrm{Re}} \frac{\mu}{h}}, \tag{30}$$

where CFL is the Courant number and h is the radius of the circumscribed circle in element I.

3.3. Numerical simulation of one-dimensional hypersonic shock structure

As the first test case, the one-dimensional hypersonic shock structure problem was considered. Since the wall boundary condition is not present in the problem, the inherent behavior of the numerical method free from the contamination caused by the solid wall boundary condition can be investigated. The shock density thickness is known as one of important parameters to assess the accuracy of the computational models in the shock structure problem. Various solutions including the second-order constitutive model [11, 20] are compared with experimental data in **Figure 2**. For better comparison, the analytic solutions of the shock density thickness recently derived by Myong [23] are also reproduced in **Figure 3**.

It can be confirmed from **Figure 2** of monatomic gases that the second-order result is in better agreement with the experimental data than the Navier-Stokes-Fourier results, implying the essential role of the second-order constitutive equation. Further, it can be found that the Eu's (unbalanced) quasi-linear generalized hydrodynamics model [3, 5] derived by ignoring the

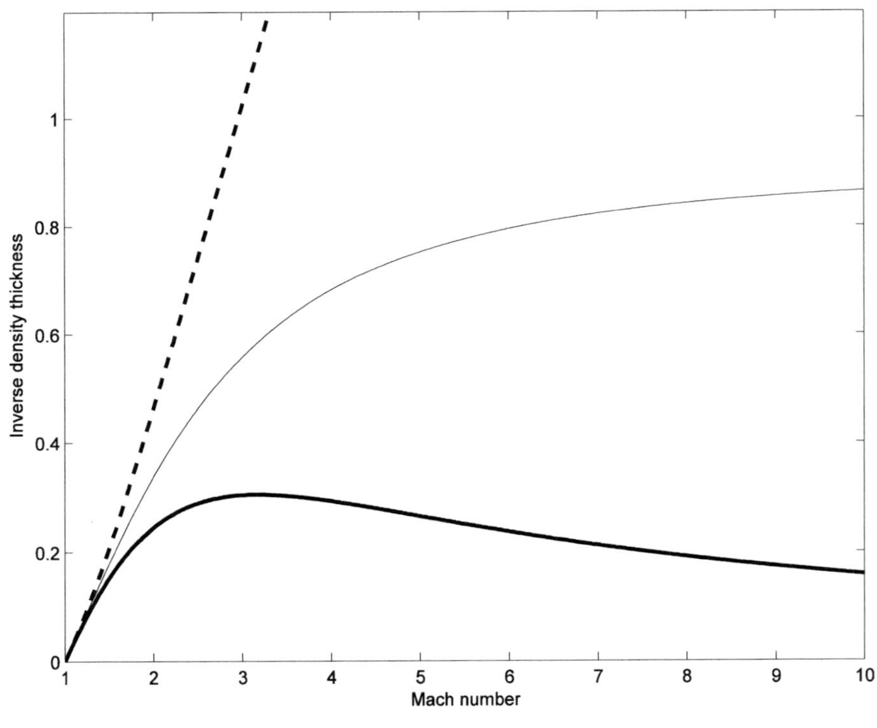

Figure 3. Inverse shock density thickness. Maxwellian molecule for the thick solid curve; hard sphere for the thin solid curve; constant case for the broken curve. (Reprinted by permission of the American Institute of Aeronautics and Astronautics, Inc. [23]).

stress-strain coupling term $2[\mathbf{\Pi} \cdot \nabla \mathbf{u}]^{(2)}$ of quadratic nature while keeping q_{2nd} predicts most poorly, even worse than the Navier-Stokes-Fourier constitutive equation does. This in turn implies that the balancing treatment plays a critical role in the closure theory.

3.4. Numerical simulations of multi-dimensional hypersonic rarefied flows

As the second test case, the two-dimensional hypersonic rarefied flows past a circular cylinder were considered [16, 26]. The input parameters for this hypersonic case are $M = 5.48$, $p = 5$ Pa, $T = 26.6$ K for far-field, and $T = 293.15$ K for solid wall. Working monatomic gas is assumed argon with Pr = 2/3. The Langmuir slip and jump boundary conditions [12, 13, 27] are applied at the solid surface. The results of both the first-order NSF and the second-order nonlinear

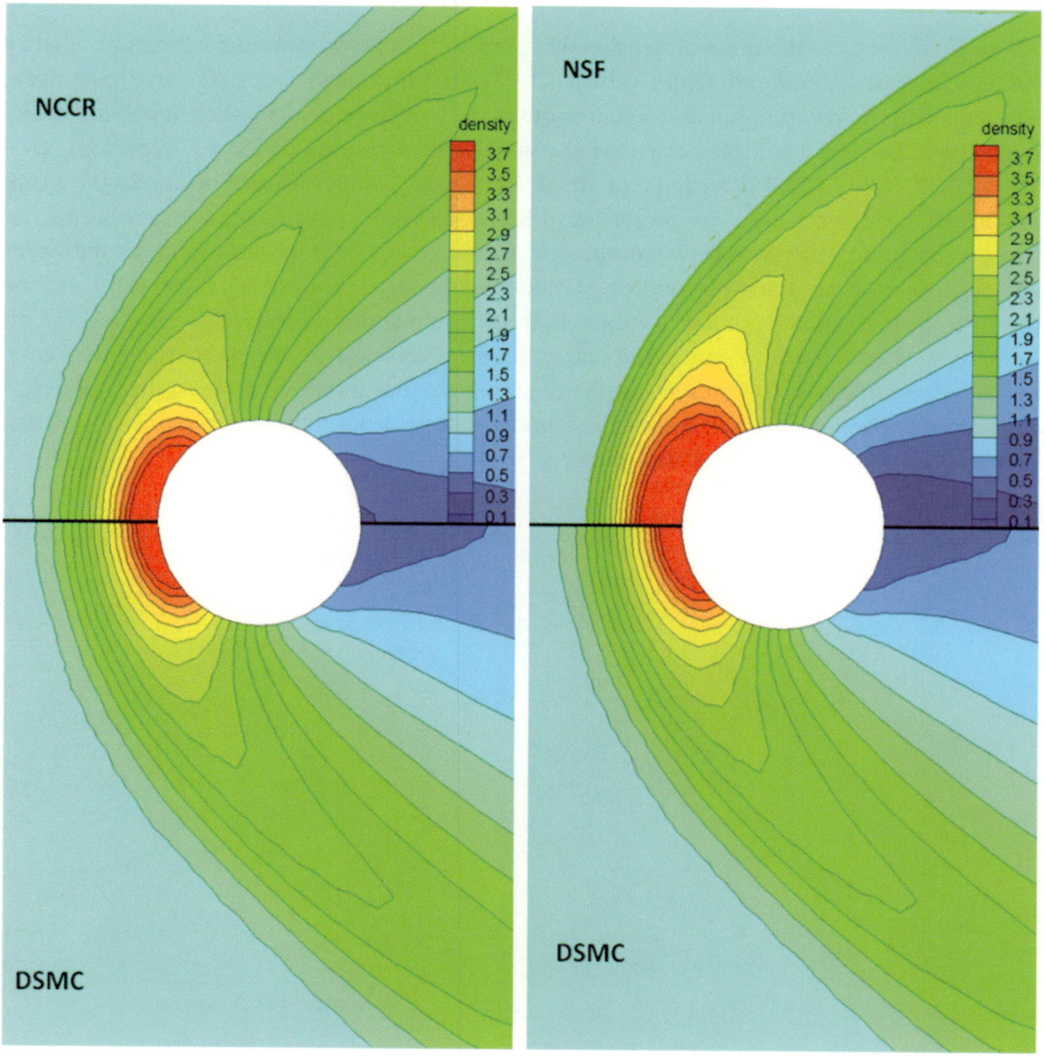

Figure 4. Normalized density fields and contours of the two-dimensional hypersonic gas flows past a circular cylinder, $M = 5.48$ and Kn = 0.5. (Reprinted with permission from Elsevier [16]).

coupled constitutive relation (NCCR) models are compared with DSMC data, which are generated by assuming full tangential momentum and thermal accommodation for slip and jump boundary conditions.

Detailed comparisons of normalized density contours of hypersonic rarefied case Kn = 0.5 [16] are presented in **Figure 4**. The results of the case Kn = 0.5 show that the density contours and the stand-off shock structure predicted by the NCCR model and the DSMC are in excellent agreement, even in this high transitional regime. On the other hand, the thickness of stand-off shock structure predicted by the first-order NSF model is much smaller than that of the second-order NCCR model and DSMC. In addition, the degree of gaseous expansion near the rear part of the cylinder predicted by the NSF model is considerably higher than that of the NCCR model and DSMC. On the whole, the results of the second-order NCCR model show better agreement with DSMC data than the first-order NSF results in hypersonic rarefied cases studied.

As the final test case, the three-dimensional hypersonic gas flows around a suborbital re-entry vehicle, Intermediate eXperimental Vehicle (IXV) of the European Space Agency (ESA), were investigated. The computational domain is defined by unstructured meshes; tetrahedron elements of 978,445 in this three-dimensional case. The flow conditions for the hypersonic case are M = 5.0, Kn = 0.02, and an angle of attack 15 degree. Comparisons of normalized density and Mach number contours are presented in **Figures 5** and **6**. On the whole, there seems not much substantial difference between numerical solutions of the first-order and second-order constitutive models, since the degree of nonequilibrium is not high. However, it can be observed from the Mach number contours that some nonequilibrium effects begin to show up in the bow shock structure and in the rear part of the vehicle where rapid expansion occurs. Besides these findings, the present results demonstrate that the three-dimensional numerical simulation of the second-order constitutive model is possible for hypersonic rarefied flows like re-entry vehicles with complicated geometry.

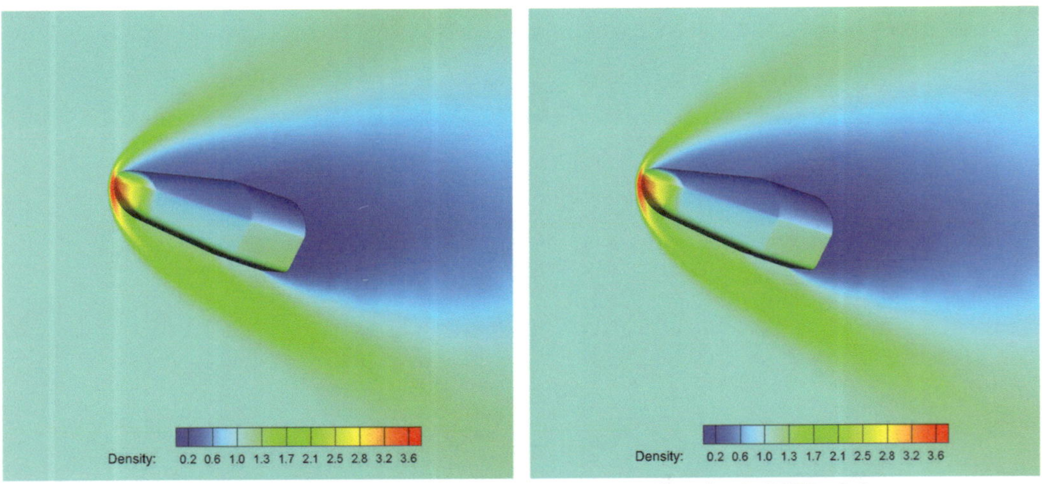

First-order NSF solution Second-order NCCR solution

Figure 5. Normalized density fields and contours of the three-dimensional hypersonic gas flows around a suborbital re-entry vehicle, M = 5.0 and Kn = 0.02.

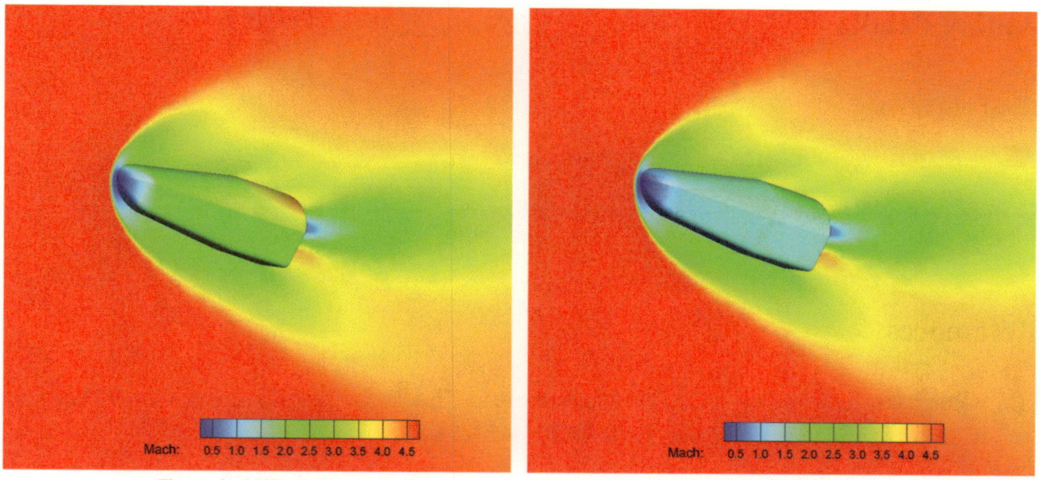

First-order NSF solution   Second-order NCCR solution

Figure 6. Mach number fields and contours of the three-dimensional hypersonic gas flows around a suborbital re-entry vehicle, $M = 5.0$ and $\mathrm{Kn} = 0.02$.

4. Conclusions

A systematic derivation of the second-order constitutive equations from the kinetic Boltzmann equation is presented. The core frameworks employed in developing the thermodynamically-consistent constitutive models are a modified moment method, called Eu's generalized hydrodynamics, and the new closure theory, called balanced closure, recently developed by Myong. Then, multi-dimensional computational models of the second-order constitutive equations are developed. The core concepts used in developing the models are the decomposition and the method of iterations. Further, as the basic computational scheme to efficiently solve the conservation laws together with the second-order constitutive equations, a mixed explicit modal DG method is developed. In order to assess the potential of the new computational model in hypersonic flow regimes, several flow problems, including the one-dimensional shock structure and three-dimensional hypersonic gas flows around a suborbital IXV re-entry vehicle, are numerically simulated. On the whole, the new second-order model is found to enhance considerably the prediction capability of hypersonic rarefied flows in comparison with the conventional first-order model.

Acknowledgements

This work was supported by the National Research Foundation of Korea funded by the Ministry of Science and ICT (NRF 2017-R1A2B2007634), South Korea. The author gratefully acknowledges the contributions of graduate students S. Singh, P. Raj, and A. Karchani for producing **Figures 2**, **5**, and **6**.

Author details

Rho Shin Myong

Address all correspondence to: myong@gnu.ac.kr

Gyeongsang National University, Jinju, South Korea

References

[1] Chapman S, Cowling TG. The Mathematical Theory of Non-Uniform Gases. London: Cambridge University Press; 1970. DOI: 10.1021/ed018p48.2

[2] Eu BC. A modified moment method and irreversible thermodynamics. Journal of Chemical Physics. 1980;**73**:2958-2969. DOI: 10.1063/1.440469

[3] Eu BC. Kinetic Theory and Irreversible Thermodynamics. New York: Wiley; 1992

[4] Eu BC. Nonequilibrium Statistical Mechanics: Ensemble Method. Dordrecht: Kluwer Academic Publishers; 1998

[5] Eu BC. Kinetic Theory of Nonequilibrium Ensembles, Irreversible Thermodynamics, and Generalized Hydrodynamics: Volume 1. Nonrelativistic Theories. Switzerland: Springer; 2016

[6] Truesdell C, Muncaster RG. Fundamentals of Maxwell's Kinetic Theory of a Simple Monatomic Gas. New York: Academic Press; 1980

[7] Bird GA. Molecular Gas Dynamics and the Direct Simulation of Gas Flows. Clarendon: Oxford Science; 1994

[8] Grad H. On the kinetic theory of rarefied gases. Communications on Pure and Applied Mathematics. 1949;**2**(4):331-407. DOI: 10.1002/cpa.3160020403

[9] Grad H. The profile of a steady plane shock wave. Communications on Pure and Applied Mathematics. 1952;**5**(3):257-300. DOI: 10.1002/cpa.3160050304

[10] Agarwal RK, Yun KY, Balakrishnan R. Beyond Navier–Stokes: Burnett equations for flows in the continuum–transition regime. Physics of Fluids. 2001;**13**:3061-3085. DOI: 10.1063/1.1397256

[11] Myong RS. Thermodynamically consistent hydrodynamic computational models for high-Knudsen-number gas flows. Physics of Fluids. 1999;**11**:2788-2802. DOI: 10.1063/1.870137

[12] Myong RS. A computational method for Eu's generalized hydrodynamic equations of rarefied and microscale gasdynamics. Journal of Computational Physics. 2001;**168**:47-72. DOI: 10.1006/jcph.2000.6678

[13] Myong RS. A generalized hydrodynamic computational model for rarefied and microscale diatomic gas flows. Journal of Computational Physics. 2004;**195**:655-676. DOI: 10.1016/j.jcp.2003.10.015

[14] Ahn JW, Kim C. An axisymmetric computational model for generalized hydrodynamic theory of rarefied and multi-species gas flows. Journal of Computational Physics. 2009;**228**:4088-4117. DOI: 10.1016/j.jcp.2009.02.026

[15] Myong RS. On the high Mach number shock structure singularity caused by overreach of Maxwellian molecules. Physics of Fluids. 2014;**26**(5):056102. DOI: 10.1063/1.4875587

[16] Le NTP, Xiao H, Myong RS. A triangular discontinuous Galerkin method for non-Newtonian implicit constitutive models of rarefied and microscale gases. Journal of Computational Physics. 2014;**273**:160-184. DOI: 10.1016/j.jcp.2014.05.013

[17] Xiao H, Myong RS. Computational simulations of microscale shock–vortex interaction using a mixed discontinuous Galerkin method. Computers & Fluids. 2014;**105**:179-193. DOI: 10.1016/j.compfluid.2014.09.027

[18] Rana A, Ravichandran R, Park JH, Myong RS. Microscopic molecular dynamics characterization of the second-order non-Navier-Fourier constitutive laws in the Poiseuille gas flow. Physics of Fluids. 2016;**28**(8):082003. DOI: 10.1063/1.4959202

[19] Myong RS. Theoretical description of the gaseous Knudsen layer in Couette flow based on the second-order constitutive and slip-jump models. Physics of Fluids. 2016;**28**(1): 012002. DOI: 10.1063/1.4938240

[20] Karchani A. Discontinuous Galerkin Methods for the Second-order Boltzmann-based Hydrodynamic Models [Ph.D. thesis]. South Korea: Gyeongsang National University; 2017

[21] Alsmeyer H. Density profiles in argon and nitrogen shock waves measured by the absorption of an electron beam. Journal of Fluid Mechanics. 1976;**74**:497-513. DOI: 10.1017/S0022112076001912

[22] Al-Ghoul M, Eu BC. Generalized hydrodynamics and shock waves. Physical Review E. 1997;**56**:2981. DOI: 10.1103/PhysRevE.56.2981

[23] Myong RS. Analytical solutions of shock structure thickness and asymmetry within Navier-Stokes/Fourier framework. AIAA Journal. 2014;**52**(5):1075-1080. DOI: 10.2514/1.J052583

[24] Bassi F, Rebay S. A high-order accurate discontinuous finite element method for the numerical solution of the compressible Navier-Stokes equations. Journal of Computational Physics. 1997;**131**:267-279. DOI: 10.1006/jcph.1996.5572

[25] Cockburn B, Shu CW. Runge-Kutta discontinuous Galerkin methods for convection dominated problems. Journal of Scientific Computing. 2001;**16**:173-261. DOI: 10.1023/A:1012873910884

[26] Bird GA, Broadwell JE, Rungaldier H, Vogenitz FW. Theoretical and experimental study of rarefied supersonic flows about several simple shapes. AIAA Journal. 1968;**6**:2388-2394. DOI: 10.2514/3.4999

[27] Myong RS. Gaseous slip model based on the Langmuir adsorption isotherm. Physics of Fluids. 2004;**16**(1):104-117. DOI: 10.1063/1.1630799

Chapter 2

Numerical Study of Hypersonic Boundary Layer Receptivity Characteristics Due to Freestream Pulse Waves

Xiaojun Tang, Juan Yu, Tianli Hui,
Fenglong Yang and Wentao Yu

Additional information is available at the end of the chapter

http://dx.doi.org/10.5772/intechopen.70660

Abstract

A finite difference method is used to do direct numerical simulation (DNS) of hypersonic unsteady flowfield under the action of freestream pulse wave. The response of the hypersonic flowfield to freestream pulse wave is studied, and the generation and evolution characteristics of the boundary layer disturbance waves are discussed. The effects of the pulse wave types on the disturbance mode in the boundary layer are investigated. Results show that the freestream disturbance waves significantly change the shock standoff distance, the distribution of flowfield parameters and the thermodynamic state of boundary layer. In the nose area, the main disturbance modes in the boundary layer are distributed near the fundamental mode. With the evolution of disturbance along with streamwise, the main disturbance modes are transformed from the dominant state of the fundamental mode to the collective leadership state of the second order and the third order harmonic frequency. The intensity of bow shock has significant effects on both the fundamental mode and the harmonic modes in each order. The strong shear structure of boundary layer under different types of freestream pulse waves reveals different stability characteristics. The effects of different types of freestream pulse waves are significant on the distribution and evolution of disturbance modes. The narrowing of frequency band and the decreasing of main disturbance mode clusters exist in the boundary layer both for fast acoustic wave, slow acoustic wave and entropy wave.

Keywords: hypersonic flow, boundary layer, receptivity, pulse wave, numerical simulation

1. Introduction

With the progress that human achieved in the exploration of aviation and aerospace fields, the hypersonic vehicle technology has developed quickly. The design of hypersonic vehicle is a complex problem based on hypersonic aerodynamics and involving multidisciplinary and multi-domain. Problems of hypersonic aerodynamics and aerodynamic heating, structures and materials technology are the main technological difficulties. There remains much variance between hypersonic flow and supersonic flow about the problem of aerodynamics. For the hypersonic boundary layer, entropy layer and shock wave layer overlapping each other, while it is high temperature and low density flow in boundary layer, which undoubtedly makes the hypersonic boundary layer flow more complex. Therefore, hypersonic aerodynamics is a key technology in the research and development of hypersonic vehicle. Various flow disturbance problems, usually exists in the flying environment of vehicle. For instance, the explosive blast wave [1], reverse jet [2, 3], the non-uniformity flow, the instability of flight, the rough wall and so on, which would happen in flying. It can be seen disturbances are common in flow fields. It's significantly different about the ideal hypothesis state of the steady flowfield and the flow condition existing in disturbance waves. The disturbance in the flowfield will have a significant influence on aerothermodynamics characteristics. The disturbance, whether it is strong or weak, will interfere with the flowfield, especially the shock and boundary layers. After the disturbance in the flowfield interference with shock wave and boundary layer, the disturbance wave will be induced. Then the induced wave will cause further interaction with boundary layer, and create new unstable waves. The stability characteristic of boundary layer and the laminar-turbulent transition mechanism will be significantly changed due to the induced unstable waves. Laminar-turbulent transition not only affects the aerodynamic heating of the wall of hypersonic vehicle, but directly changes the aerodynamic force. Especially, laminar-turbulent transition will greatly increases aerodynamic drag, which reduces the lift drag ratio of hypersonic vehicle significantly and increases the requirement of thermal protection.

Therefore, it is necessary to accurately predict the hypersonic unstable flowfield and the flowfield response characteristic induced disturbance waves during the design and development process of hypersonic aircraft. Considering the complexity and expensiveness of the hypersonic vehicle wind tunnel test, it is of practical significance to conduct the numerical simulation of hypersonic unsteady flowfield. In recent years, the hypersonic flowfield response induced by different disturbance waves and the influences of the disturbance wave on the stability of the boundary layer are studied by many scholars using numerical or experimental methods. Ma and Zhong [4] investigated the response of hypersonic boundary layer over a blunt cone to freestream acoustic waves at Mach 7.99. The receptivity of a flat plate boundary layer to a freestream axial vortex is discussed by Boiko [5]. Zhong [6] investigated the leading-edge receptivity to freestream disturbance waves for hypersonic flow over a parabola. The effect of wall disturbances on hypersonic flowfield, and the response of hypersonic boundary layer to wall disturbances are also widely studied [7–10]. Literature [11] points out that, after the interaction between any form of freestream disturbance and the shock wave as well as the boundary layer in hypersonic flow field, three independent forms of disturbance waves, including acoustic disturbance (fast and slow acoustic disturbance), entropy wave disturbance, and vortex wave disturbance will be generated. Among these investigations, most scholars are committed to study the effects of freestream continuous disturbance or the effects of wall disturbance on the stability of boundary layer and laminar-turbulent transition [12–14]. There is still less research on the effects of freestream pulse disturbance on the stability of boundary layer, and the mechanism in this field is still not fully understood. So, this paper aims to study hypersonic flowfield and the stability characteristics of boundary layer under the freestream pulse wave. It should be mentioned that the significantly different showed in the influence mechanism of the wave types of continuous disturbance on the stability of boundary layer that had been pointed out by some scholars [6, 15, 16]. Ma and Zhong [15, 16] investigated the response of hypersonic plate boundary layer under different types of disturbance waves using direct numerical simulation and linear stability theory. It is found that acoustic disturbance has the greatest influence on the stability of boundary layer, and it is pointed out that the action mechanism of freestream entropy wave disturbance and vortex wave disturbance is similar to that of the fast acoustic wave disturbance, but different from the slow wave disturbance. Although it is significantly different between pulse disturbance wave and continuous disturbance wave in form, there is reason to believe that the response characteristics of hypersonic boundary layer are also significantly different under different freestream pulse waves. Therefore, the receptivity characteristics of hypersonic boundary layer under freestream pulse waves with different types are discussed in this paper.

In this investigation, the numerical simulations of hypersonic flowfield over a blunt wedge under the action of freestream pulse wave were conducted using a high order finite difference scheme. The response property of hypersonic flowfield under the action of freestream pulse wave analyzed, and the hypersonic boundary layer stability is investigated. The receptivity characteristic of hypersonic boundary layer under the action of different freestream pulse waves was compared, and the effects of the pulse wave types on the generation and evolution of the disturbance mode in the hypersonic boundary layer are discussed.

2. Basic equations and numerical method

2.1. Navier-Stokes equation

According to the forms of conservation equation, momentum equation and energy equation, the three basic equations of fluid governing equations can be written as a general form, that is, the two-dimensional unsteady compressive N-S equation can be expressed as:

$$\frac{\partial \mathbf{Q}}{\partial t}+\frac{\partial(\mathbf{f}\text{-}\mathbf{f}_v)}{\partial x}+\frac{\partial(\mathbf{g}\text{-}\mathbf{g}_v)}{\partial y}=0 \tag{1}$$

Where the state vector $\mathbf{Q}$ can be expressed as:

$$\mathbf{Q}=[\rho\ \rho u_x\ \rho u_y\ \rho E]^{\mathrm{T}} \tag{2}$$

Similarly, $\mathbf{f}$, $\mathbf{f_v}$, $\mathbf{g}$, $\mathbf{g_v}$ can be expressed respectively as:

$$\mathbf{f}=\left[\rho u_x \quad \rho {u_x}^2+P \quad \rho u_x u_y \quad (\rho E+P)u_x\right]^{\mathrm{T}} \tag{3}$$

$$\mathbf{f}_v=\left[0 \quad \tau_{xx} \quad \tau_{xy} \quad u_x\tau_{xx}+u_y\tau_{xy}+k\frac{\partial T}{\partial x}\right]^{\mathrm{T}} \tag{4}$$

$$\mathbf{g} = \begin{bmatrix} \rho u_y & \rho u_x u_y & \rho {u_y}^2 + p & (\rho E + p) u_y \end{bmatrix}^{\mathrm{T}} \tag{5}$$

$$\mathbf{g}_v = \begin{bmatrix} 0 & \tau_{xy} & \tau_{yy} & u_x \tau_{xy} + u_y \tau_{yy} + k \dfrac{\partial T}{\partial y} \end{bmatrix}^T \tag{6}$$

where u_x, u_y, P, ρ, T and E indicate the velocity in the x direction, the velocity in the y direction, pressure, density, temperature and total energy, respectively. k, τ and E are the thermal conductivity, the stress component and the total energy, respectively.

2.2. Numerical method

In this paper, a high order finite difference method is used to solve the flowfield governing equation directly. The inviscid vector flux of the Navier-Stokes equation is divided into the positive and negative convection terms using S-W method [17]. The positive and negative convection terms are discretized by the 5th order upwind WENO scheme [18]. The viscous term is discretized by the 6th order central difference scheme [19]. In order to obtain the transient information of the flowfield and reduce computation time, the 3th TVD Runge-Kutta scheme is used for time advance [20].

The spatial discretization for positive and negative convection terms can be expressed as Eqs. (7) and (8), respectively:

$$W' = \left(m_1 W_{j+3} + m_2 W_{j+2} + m_3 W_{j+1} + m_4 W_j + m_5 W_{j-1} + m_6 W_{j-2} + m_7 W_{j-3} + m_8 W_{j-4}\right)/\Delta \tag{7}$$

$$H' = \left(n_1 H_{j+4} + n_2 H_{j+3} + n_3 H_{j+2} + n_4 H_{j+1} + n_5 H_j + n_6 H_{j-1} + n_7 H_{j-2} + n_8 H_{j-3}\right)/\Delta \tag{8}$$

where W and H are positive convection terms and negative convection terms, respectively; Δ is the grid spacing; W' and H' are the difference approximation of the derivative of W and H, respectively; m_i and n_i are specific coefficients, no longer list.

The spatial discretizations for viscous terms can be expressed as follow:

$$L' = \left(K_1\left(L_{j+1} - L_{j-1}\right) + K_2\left(L_{j+2} - L_{j-2}\right) + K_3\left(L_{j+3} - L_{j-3}\right)\right)/\Delta \tag{9}$$

where L, L', Δ and K_i are viscous terms, the difference approximation of the derivative of viscous terms, the grid spacing and specific coefficients, respectively, which are similar to the symbols in Eqs. (7) and (8).

The TVD Runge-Kutta discretization can be expressed as follows

$$\begin{cases} U^{(1)} = a_1 U^n + b_1 f^n \Delta t & U^{(2)} = a_2 U^n + b_2\left(U^{(1)} + f^1 \Delta t\right) \\ U^{(3)} = a_3 U^n + b_3\left(U^{(2)} + f^2 \Delta t\right) & U^{n+1} = U^{(3)} \end{cases} \tag{10}$$

where Δt is time increment for time advance; L(U) is the partial derivative of U relative to time.

The variables a_i and b_i is the specific coefficients. b_1, b_2 and b_3 are equal to 1, 3/4 and 1/3, respectively; a_1, a_2 and a_3 are equal to 1, 1/4 and 2/3, respectively.

3. Computing model and conditions

The model parameters and calculation conditions of flowfield calculation includes the freestream condition, model parameter, boundary conditions and meshing. For freestream conditions, the freestream temperature is equal to 69 K, and the Mach number is equal to 6. The Reynolds number, based on the nose radius, is equal to 10,000. The angle of attack is equal to 0°. **Figure 1** shows the computing model and schematic diagram. Calculation mode is a blunt wedge with the wedge angle of 16°e; the nose radius $r = 1$ mm. the adiabatic wall, no-penetration and non-slip is introduced for wall condition; the symmetric boundary conditions is introduced for y = 0; the freestream conditions and exportation boundary conditions are introduced for the upstream of computing domain and the downstream of calculation domain, respectively. **Figure 2** shows the computational grid. The local intensive grid method is carried out near the nose area and wall area, and the grid number is 300 × 120. The parameters of blunt wedge nose radius r, freestream velocity V_∞, freestream temperature T_∞, freestream viscosity coefficient μ_∞, freestream conductivity coefficient k_∞, freestream density ρ_∞ are chosen as the normalized basic measure that is the characteristic variable.

In order to explore the influence of freestream pulse wave on hypersonic flow field, the interaction process between freestream pulse slow acoustic wave and hypersonic flowfield is direct numerical simulated. In present investigation, the stable flow over a blunt wedge at Mach 6 is calculated firstly, and then the simulation of hypersonic unsteady flowfield under the freestream pulse wave is conducted by introducing freestream pulse slow acoustic wave at the upstream boundary of computing domain.

The introducing time of the freestream pulse wave is recorded as $t = 0$. The form of pulse slow acoustic wave is expressed as follows:

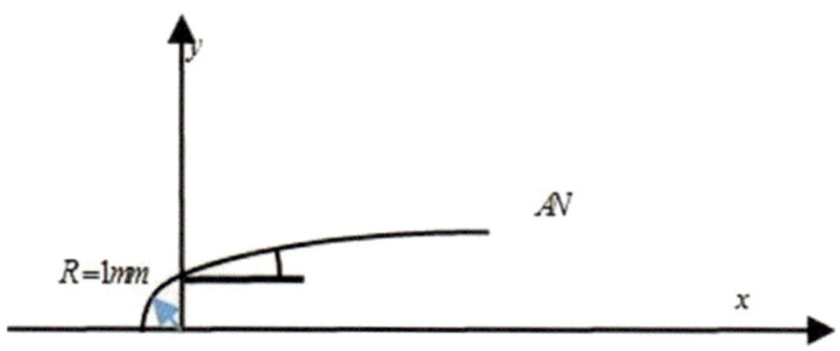

Figure 1. Computational model and schematic diagram.

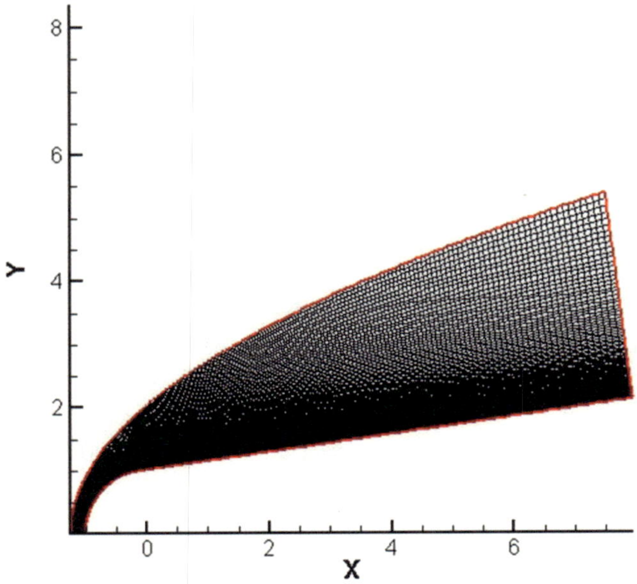

Figure 2. Mesh grid.

$$\begin{bmatrix} u' \\ v' \\ p' \\ \rho' \end{bmatrix} = \begin{cases} [0\ 0\ 0\ 0]^T & (t<0) \\ \left[A\ \ 0\ \ -\frac{A}{M_\infty}\ \ -AM_\infty\right]^T e^{i\left(kx-\frac{FRe}{10^6}t+\frac{\pi}{2}\right)} & (0\le t<2) \\ [0\ \ 0\ \ 0\ \ 0]^T & (t\ge 2) \end{cases} \tag{11}$$

Where u', v', P' and ρ' indicate the velocity disturbance along x direction, the velocity disturbance along y direction, the pressure disturbance and the density disturbance, respectively; A indicates the amplitude, F indicates the generalized frequency, t indicates time. Here, $A = 8 \times 10^{-2}$; $k = 3.1446 \times 10^{-4}$; $F = 50\pi$; $Ma_\infty = 6$.

4. Results and discussion

4.1. Response of the hypersonic flowfield to freestream pulse wave

4.1.1. Response of the hypersonic flowfield

Before analyzing, it should be pointed out that the numerical simulation method and the grid independence in present work is validated in our previous investigation [21–23], which indicates the numerical method adopted in this paper is reliable. **Figure 5** shows the contours of pressure under freestream pulse wave at different times.

Figure 3(a)–(d) in the figure correspond to t = 2.0, 4.0, 6.0 and 8.0, respectively. In **Figure 3**, when the pulse wave enters the flow field, it first interacts with the bow-shaped shock wave. When it is encountered with the slow acoustic pulse wave, the bow-shaped shock wave is

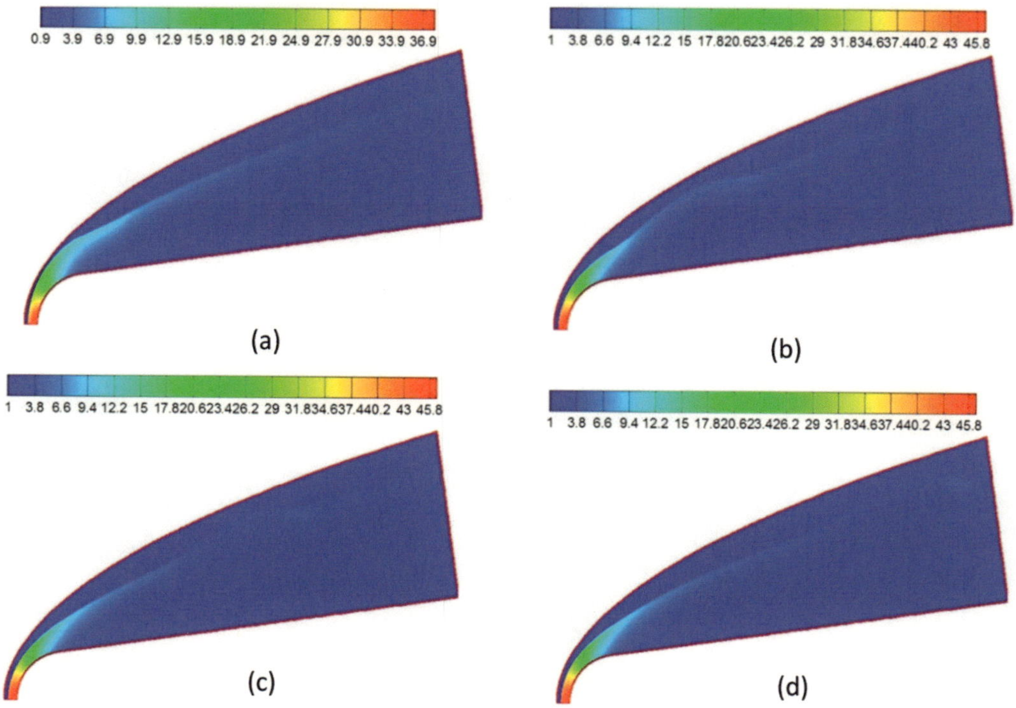

Figure 3. Contours of pressure under freestream pulse wave at different times. (a) t=2.0, (b)) t=4.0, (c)t=6.0, (d)) t=8.0.

deformed and protruded outward, and the surrounding pressure is also significantly affected. Due to the deformation of the shock wave, the pressure of outward convex region near shock wave is significantly increased, and the flowfield pressure below the bow shock wave is greatly reduced. That is to say, from **Figure 3(a)–(d)**, it shows that the freestream disturbance waves significantly interact with the bow-shaped shock waves, which greatly changes the shock standoff distance and the distribution of flowfield pressure in the active region.

Figure 4 shows the contours of density under freestream pulse wave at different times. **Figure 4(a)–(f)** correspond to t = 0.0, 2.0, 4.0, 6.0, 8.0 and 10.0, respectively. **Figure 5** shows the contours of temperature under freestream pulse wave at different times. **Figure 5(a)–(f)** correspond to t = 0.0, 2.0, 4.0, 6.0, 8.0 and 10.0, respectively. As can be seen from **Figure 4**, the pulse wave disturbance has a great impact on the density. The density of the flowfield changes significantly under the action of freestream pulse wave, especially in the disturbance area between the pulse wave and the bow-shaped shock wave, where the density of the region is significantly smaller. Compared with the significant changes of flowfield density around the shock wave, the pulse wave has much smaller effect on the density in the boundary layer. Meanwhile, the closer to the nasal area, the greater the density changes and the effects are after the action of pulse wave. Obviously, this is because the closer the bow-shaped shock wave is to the nose area, the stronger the action is. Therefore, it can be concluded that the stronger the intensity of the shock wave is, the more intense the effect of the pulse slow acoustic wave and the bow-shaped shock wave are. As can be seen from

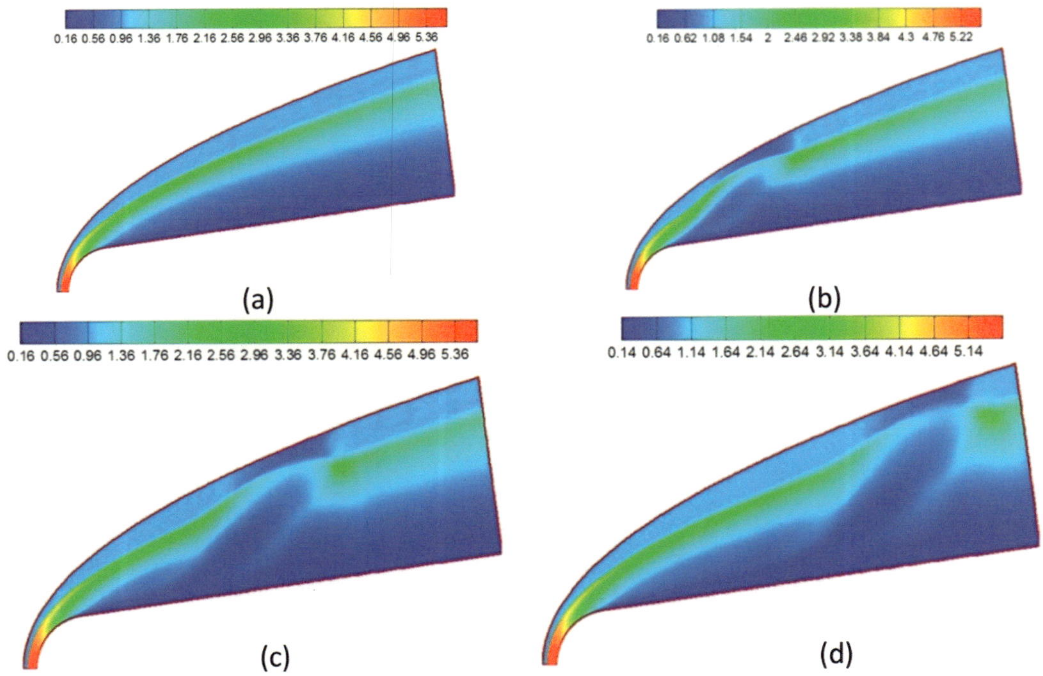

Figure 4. Contours of density under freestream pulse wave at different times. (a) t=0.0, (b) t=4.0, (c) t=6.0, (d) t=8.0.

Figure 5, the effect of disturbances on the temperature is pretty significant, which similar to the density. Under the action of impulse disturbances, a strip area exists in the hypersonic flowfield temperature disturbances between the boundary layer and the shock wave, which is similar to the change trend of the flowfield density's disturbances. What is worth noting is that from the upstream to the downstream, the strip area becomes longer; this is due to the fact that through the shock wave, the disturbance wave slows down. The farther it is away from the wall, the greater the flow velocity is, which leads to the disturbance wave's different spread velocity to the downstream. It also can be seen from **Figure 5** that the temperature in the boundary layer is much more significantly affected by the pulse wave compared with the temperature change of the flowfield near the shock wave. That is to say, the freestream pulse wave's influence on the thermodynamic state in boundary layer is significantly greater than that on the thermodynamic state the outside boundary layer. It is known that the temperature is a characteristic parameter of the thermodynamic properties in the boundary layer. Obviously, the freestream pulse disturbance wave changes the thermal mechanism of the strong shear flow in the boundary layer, while the thermodynamic mechanism is considered to have an important effect on the stability of the boundary layer [24], so it is reasonable to believe that the pulse slow acoustic wave in freestream plays a considerable role in the stability of hypersonic boundary layer.

Figure 6 shows the contours of velocity along axis y under freestream pulse wave at different times. From **Figure 6**, under the action of the pulse slow acoustic wave, the velocity in the interference area between pulse wave and shock wave, the flowfield between the boundary

http://dx.doi.org/10.5772/intechopen.70660

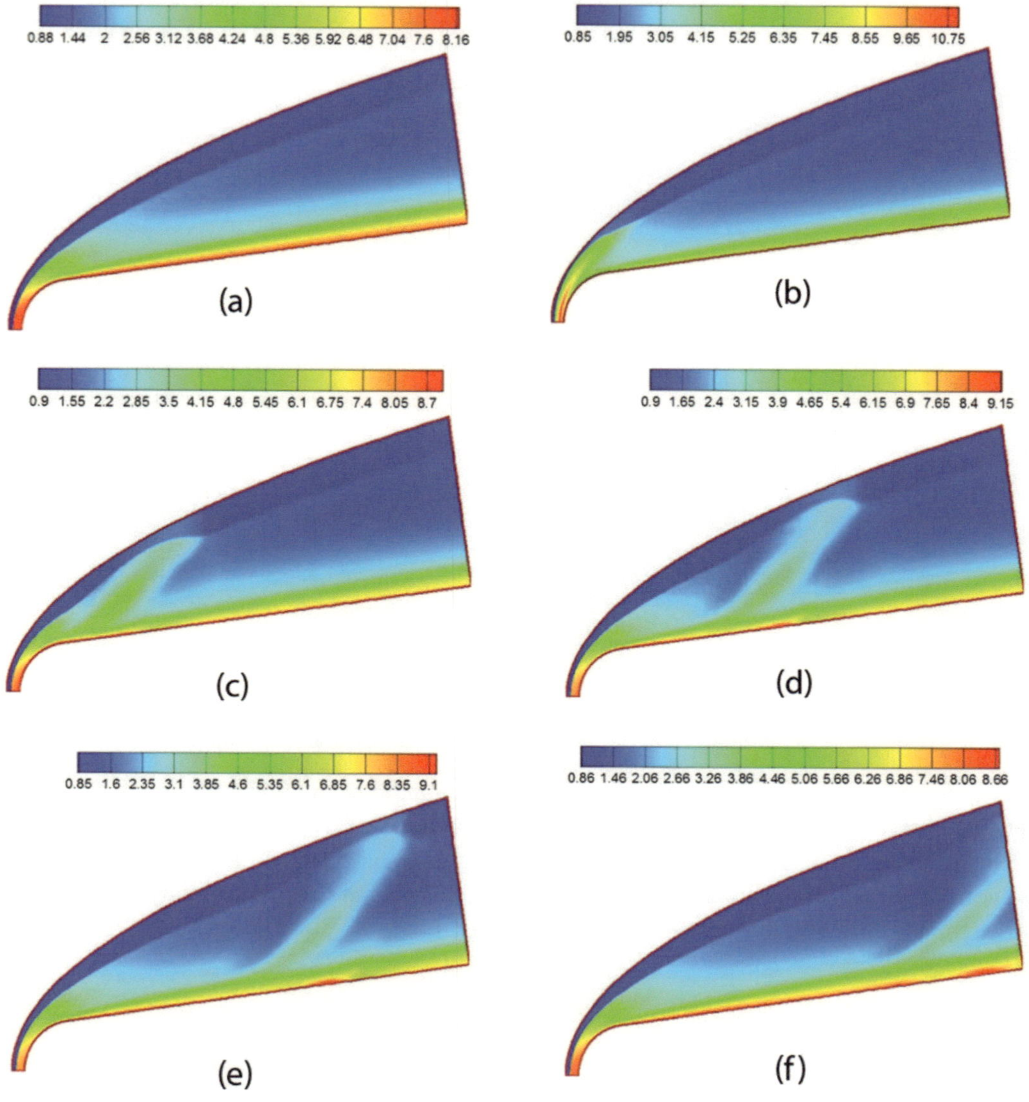

Figure 5. Contours of temperature under freestream pulse wave at different times. (a) t=0.0, (b) t=2.0, (c) t=4.0, (d) t=6.0, (e) t=8.0, (f) t=10.0.

layer and shock wave, and the flowfield in the boundary layer change greatly. On the other hand, the velocity disturbance modes out of the boundary layer and the velocity disturbance modes near the boundary layer differ significantly. From **Figure 6**, there are many obvious disturbance characteristic regions in the hypersonic boundary layer under the action of freestream pulse disturbances. It can also be seen that the boundary layer's disturbance velocity is very large, which shows that there is complex interference in the boundary layer under the action of the pulse wave. It is worth mentioning that, after the disturbance wave interfering with the bow-shaped shock wave, and entering into the flowfield, a part of the

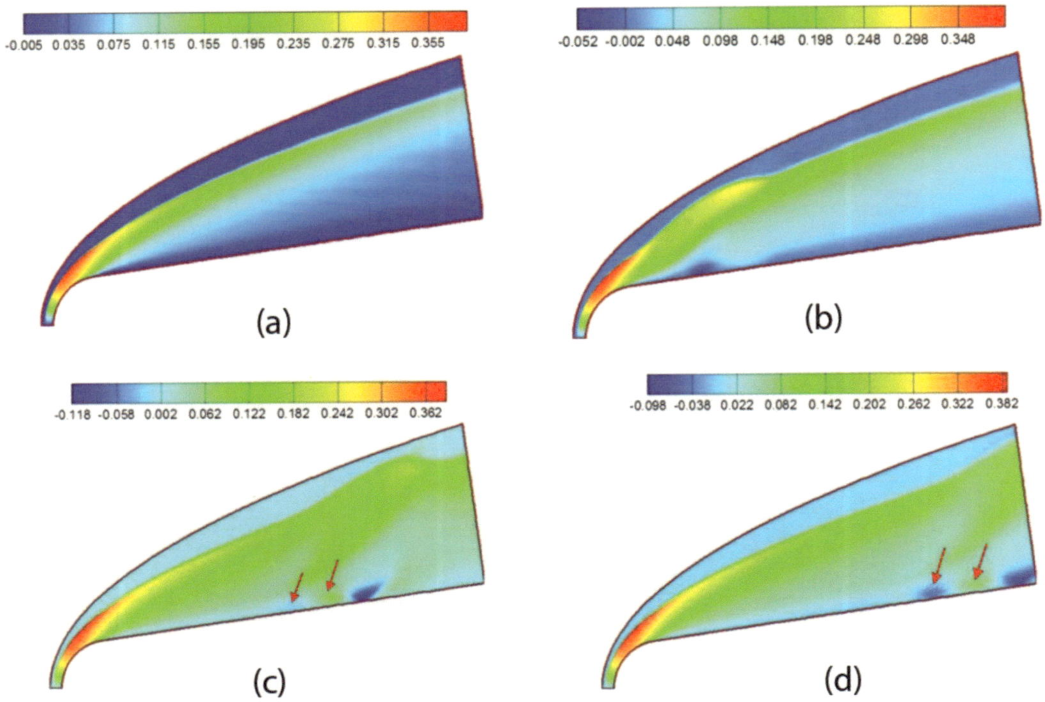

Figure 6. Contours of velocity along axis *y* under freestream pulse wave at different times. (a) t=0.0, (b) t=4.0, (c) t=8.0, (d) t=10.0.

wave will propagate from the upstream to the downstream of the flowfield and a part of the reflected wave will oscillate between the shock wave and the wall [24]. Moreover, the velocity disturbance mode in the boundary layer becomes more complicated on account of reflected waves. As shown in **Figure 6(c)** and **(d)**, the disturbance characteristic regions marked with arrows are induced by the reflected waves in the flowfield.

4.1.2. Response of the boundary layer

Figure 7 shows the distribution of friction factor disturbance on wall under freestream pulse wave at different times (t = 2.0, 4.0, 6.0, 8.0, 10.0). It can be seen from the figure that, when the pulse wave disturbance is through some place of the wall, the friction factor reduces at first, and after reaching a minimum value, it increases. When the pulse disturbance completely passes this point, the disturbance effect on this position is not over, and it will go through a number of processes in oscillating changes like getting bigger-smaller-bigger-smaller, until the change magnitude decreases to zero. It is clear that this oscillating decreasing phenomenon is caused by the reciprocating reflected waves between the wall and the bow shock. Most importantly, the friction factor is an important parameter for characterizing the shear flow, which indicates that the strong shear flow structure in the boundary layer under freestream pulse wave has changed.

Figure 8 shows the comparisons of the density, temperature and pressure disturbances on wall at different times. **Figure 8** indicates that, under freestream pulse wave, the variations of the

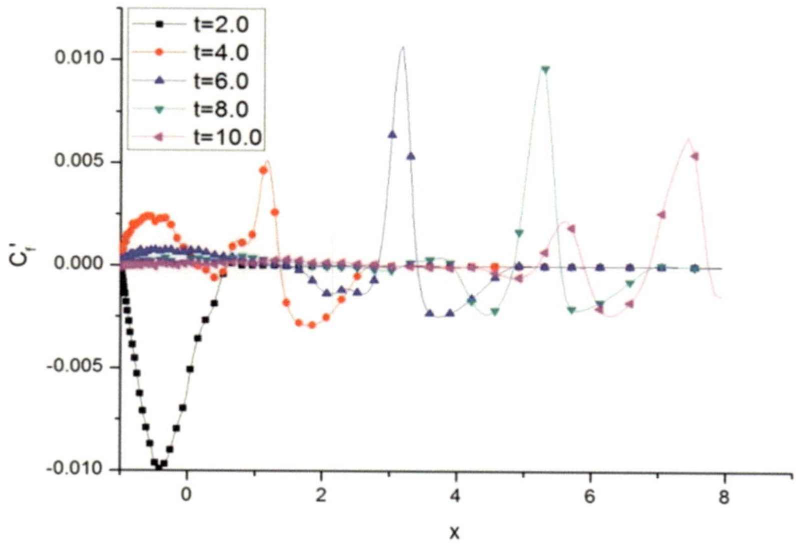

Figure 7. Distribution of friction factor disturbance on wall under freestream pulse wave at different times.

density, temperature and pressure disturbances on wall are similar to that of friction factor. Zhang et al. [19] studied the evolution of continuous small disturbance in hypersonic flow using direct numerical simulation (DNS), and found that, because of the normal shock wave, the forcing disturbance in freestream is enlarged. From **Figure 8**, it is obtained that the

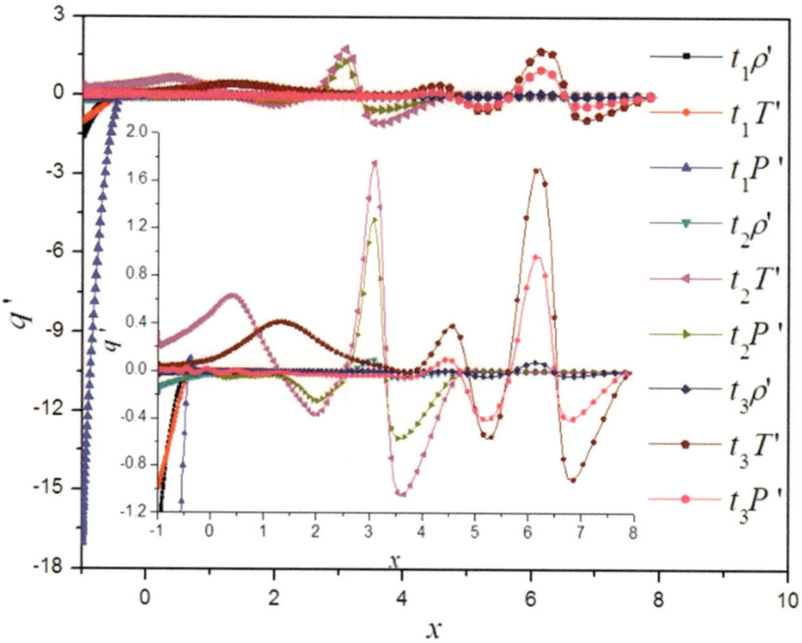

Figure 8. Comparison of the density, temperature and pressure disturbances on wall at different times ($t_1 = 1.0$, $t_2 = 6.0$, $t_3 = 9.0$).

amplitude of parameters disturbance on the wall is enlarged sharply relative to that of initial wave in freestream. The numerical results agree with Li's results. It also shows that the disturbance amplitudes of pressure on the nose of blunt wedge are larger than that on non-nose, and the former are several times, even more than 10 times, larger than the latter; while the discrepancy is tiny for temperature and density.

To study the evolution of the disturbance wave in boundary layer during the interaction process, the temperature disturbance on wall along the streamwise is considered. **Figure 9** shows the distribution of temperature disturbance on wall along the streamwise at $t = 8.6$. As discussed earlier, it has been verified that the disturbance amplitude of aerothermodynamics parameters will undergo a damped oscillation during the interaction process although there is just a half sinusoidal pulse wave in freestream. **Figure 9** indicates that the distance between crests as well as trough of disturbance variables changed markedly during the damped oscillation process. The distances a1-a2, a2-a3, b1-b2 and b2-b3 along the x-axis direction equal to 1.733, 2.939, 1.541 and 1.491, respectively. Hence, it is obvious that the damped oscillation is aperiodic. In other word, the waves with other frequencies are induced due to the interaction between freestream blast wave and bow shock wave as well as boundary layer. It should be noted that the evolution of the wave's modes in the blunt wedge boundary layer exerts an important impact on both boundary layer stability characteristic and laminar-turbulent transition [10, 25]. There is little research focused on the evolution of the wave's modes in the boundary layer under freestream pulse wave, and the evolution characteristic remains unclear although a host of in-depth researches is done in the evolution of the wave's modes in boundary layer under freestream continuous small disturbances in recent years.

To analyze the disturbance mode distribution in the hypersonic boundary layer and its evolution characteristics along the streamwise under the action of freestream pulse wave, the fast Fourier transformation method is used to decode the time domain signal of the

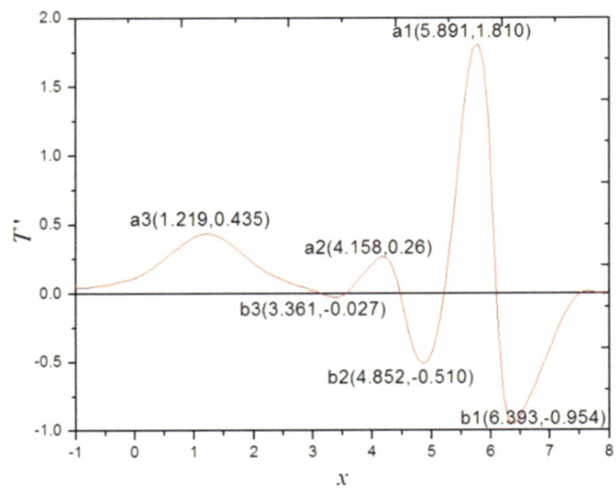

Figure 9. Temperature disturbance profile along streamwise at $t = 8.6$.

boundary layer disturbance. Furthermore, the signals are expanded in frequency order to make it as a frequency function, thereby converting the time domain signal of the pressure disturbance into a frequency domain signal. **Figure 10** shows the results of the Fourier frequency spectral analysis of pressure disturbance in boundary layer under freestream pulse wave at different locations. **Figure 10(a)–(d)** are the results of the Fourier frequency spectral analysis when $x = 0.83566$, $x = 1.2196$, $x = 3.3050$ and $x = 5.3846$, respectively. It can be seen from **Figure 10(a)** that, at the location of x = 0.83566, namely the nose area, the main disturbance modes in the boundary layer are distributed near $f = 0.25$, indicating that near the fundamental frequency, the disturbance mode component of other frequencies is relatively small, which is consistent with the results obtained by Zhang et al. [19] in studying continuous waves. It also shows that the higher the disturbance mode's frequency is, the smaller its component ratio will be. Based on **Figure 10(b)**, at the position of x = 1.2196, the disturbance mode component ratio near the fundamental frequency decreases rapidly, the high frequency disturbance harmonic mode like the second order harmonic frequency and above significantly increase. And the main disturbance modes in the boundary layer are distributed near f = 0.5, that is near the second order harmonic frequency. The main

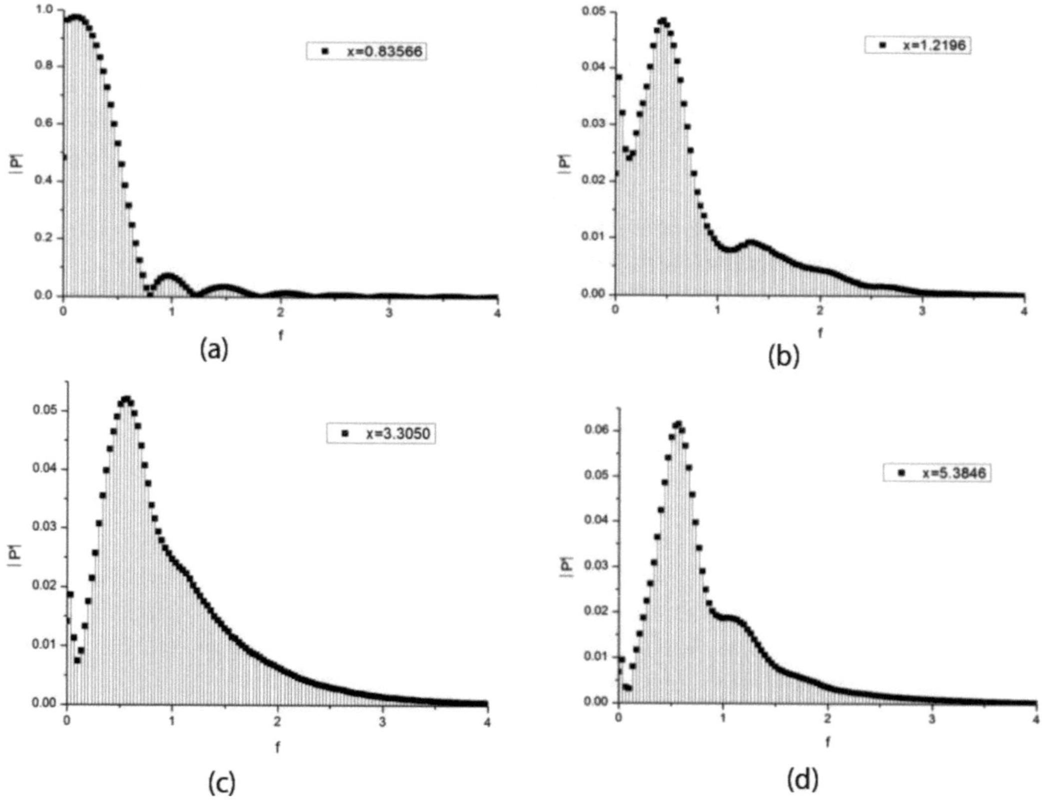

Figure 10. Fourier frequency spectral analysis of pressure disturbance in boundary layer under freestream pulse wave at different locations. (a) x=0.83566, (b)x=1.2196, (c) x=3.3050, (d) x=5.3846.

disturbance modes in the boundary layer migrate from the fundamental frequency to the second order harmonic frequency. In **Figure 10(c)**, it can be seen that at the position of $x = 3.3050$, the disturbance mode components in the vicinity of the fundamental frequency are further reduced, and the proportion of disturbance mode components is further increased. The main disturbance modes in the boundary layer are distributed at $f = 0.5–0.75$, namely near the second and third order harmonic frequency. The main disturbance modes in the boundary layer are transformed from the dominant state of the second order harmonic frequency to the collective leadership state of the second order harmonic frequency and the third order harmonic frequency. According to **Figure 10(d)**, at the position of $x = 5.3846$, the distribution of the boundary layer disturbance modes is similar to that at $x = 3.3050$. The main disturbance modes in the boundary layer are distributed near the second and third order harmonic frequency, but the frequency band of the boundary layer disturbance mode is significantly narrowed.

Figure 11 shows the evolution of different disturbance modes along streamwise in boundary layer. **Figure 11(a)–(f)** are the evolution of the fundamental frequency mode, the second order harmonic frequency, the third order harmonic frequency, the fourth order harmonic frequency, the fifth order harmonic frequency and the sixth order harmonic frequency, respectively. From **Figure 11(a)**, in the nose region, the amplitude of the fundamental frequency becomes smaller. When out of the nose region, the amplitude of the fundamental frequency gradually becomes smaller and generally remains. In **Figure 11(b)**, it can be seen that in the nose region, the amplitude of the second order harmonic frequency mode significantly becomes smaller. Out of the nose region, the amplitude of the second order harmonic frequency mode continues to increase. From **Figure 11(c)**, it can be seen that the amplitude of the third order harmonic frequency mode becomes smaller in the nose region, and the third order harmonic frequency mode first increases and then decreases in the downstream evolution. From **Figure 11(d)–(f)**, it shows that the evolution of the fourth order harmonic frequency mode, the fifth order harmonic frequency mode and the sixth order harmonic frequency mode along the streamwise is similar to that of the third order harmonic frequency mode. As shown in the figure, whether it is the fundamental mode or harmonic modes, in the area of the nose boundary layer, the amplitude of the disturbance significantly becomes smaller, indicating that the intensity of the bow shock in the hypersonic flowfield becomes weak, and it has significant effects on both the fundamental mode and the harmonic modes in each order. According to the figure, it also shows that except for the fundamental mode, from the nose area to the non-nose area, the amplitude of each harmonic frequency mode all increase significantly. It is believed that this phenomenon is caused by the recompression of the flowfield, which means in this region, the effect of the shock wave intensity on the harmonic frequency modes of the boundary layer is smaller than the recompression effect, and the effect of the shock wave intensity on the fundamental mode of the boundary layer is larger than the recompression effect.

4.2. Effects of pulse wave type

In this section, to discuss the effects of the pulse wave types on the generation and evolution of the disturbance mode in the hypersonic boundary layer, the numerical simulations of hypersonic flowfield under the action of freestream pulse fast acoustic wave, slow acoustic wave and

http://dx.doi.org/10.5772/intechopen.70660

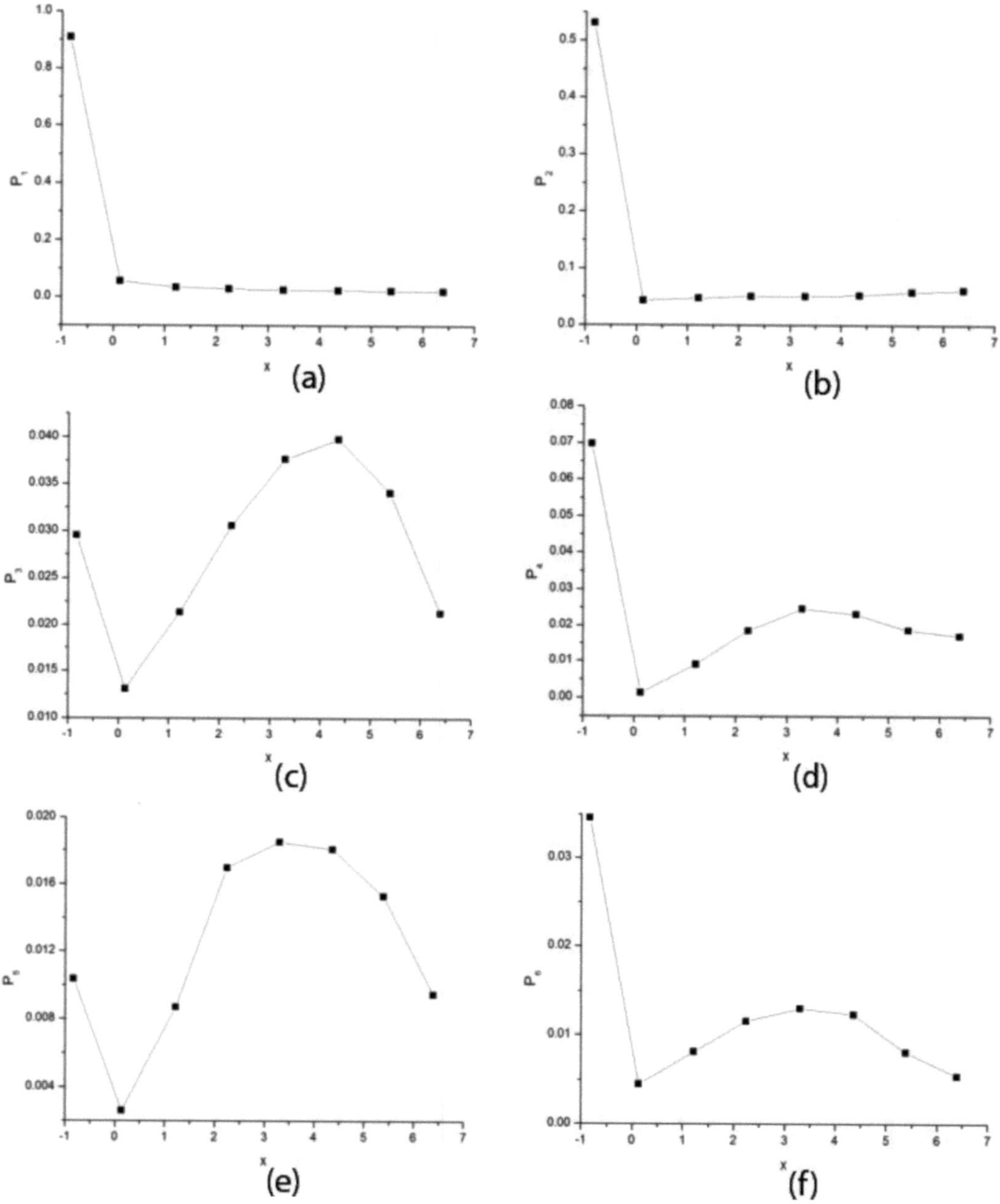

Figure 11. Evolution of different disturbance modes along streamwise in boundary layer. (a) P_1, (b) P_2, (c) P_3, (d) P_4, (e) P_5, (f) P_6.

entropy wave were conducted. It should be noted that the wall temperature $T_w = 200$ K, and the amplitude $A = 6 \times 10^{-2}$. The other computational condition is consistent with Section 4. The forms of pulse fast acoustic wave and entropy wave is no longer listed in this section. Due to the fact that the wall of shear force is decided by the friction factor on wall directly which is a key parameters for representing the strong shear structure in boundary layer, the stability of the hypersonic boundary layer is the study of the stability of the strong shear structure. **Figure 12** shows the distribution of friction factor disturbance on wall under different freestream pulse waves at $t = 6$. From **Figure 12**, it is obtained that, under freestream pulse waves, the mainstream disturbance wave influence on the wall friction coefficient under the pulse slow acoustic wave is opposite to that under pulse fast acoustic wave and entropy wave. There is significant difference between the disturbance amplitude of the wall friction coefficient under pulse fast acoustic wave and that under entropy wave. It also can be seen that there is a significant difference in the friction factor disturbance on wall caused by the reflected wave for these three cases, and the reflected wave has a significant effect on the friction factor. In general, different types of freestream pulse waves have different effects on the friction factor on wall, which indicates the interaction between the strong shear structure of boundary layer and different types of freestream pulse waves have different mechanisms of action. That is, the strong shear structure of boundary layer under the action of different types of freestream pulse waves reveals different stability characteristics.

Figure 13 shows the pressure disturbance in boundary layer under different freestream pulse waves. **Figure 13(a)** and **(b)** are corresponding the position $s = 0.92619$ and 5.90498, respectively. From **Figure 13**, it can be seen that the pressure disturbance in boundary layer under the action of different types of freestream pulse waves appear different changes in the time domain.

The Fourier frequency spectral analysis of pressure disturbance in boundary layer under different freestream pulse waves is shown in **Figure 14**. **Figure 14(a)–(c)** are corresponding the position $s = 0.92619$, $s = 3.83551$ and $s = 8.02802$, respectively. **Figure 14** shows that the

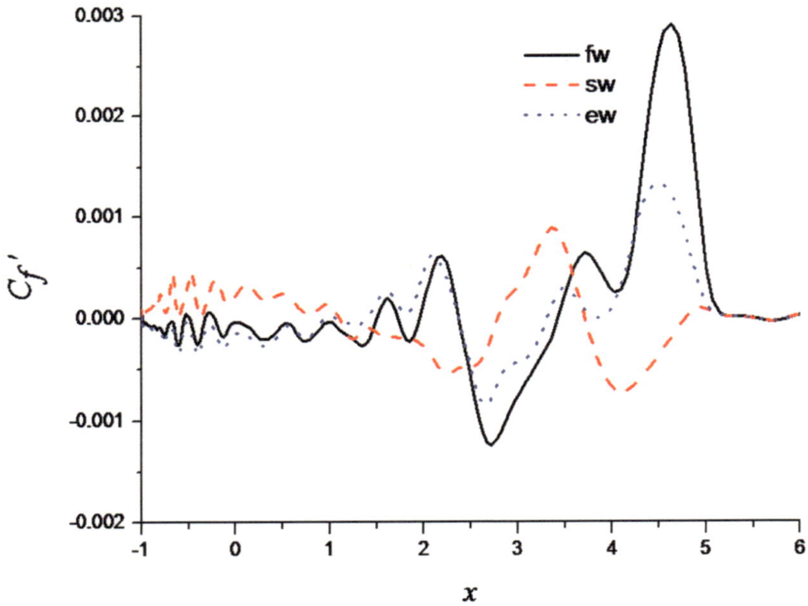

Figure 12. Distribution of friction factor disturbance on wall under different freestream pulse waves at t = 6 s.

http://dx.doi.org/10.5772/intechopen.70660

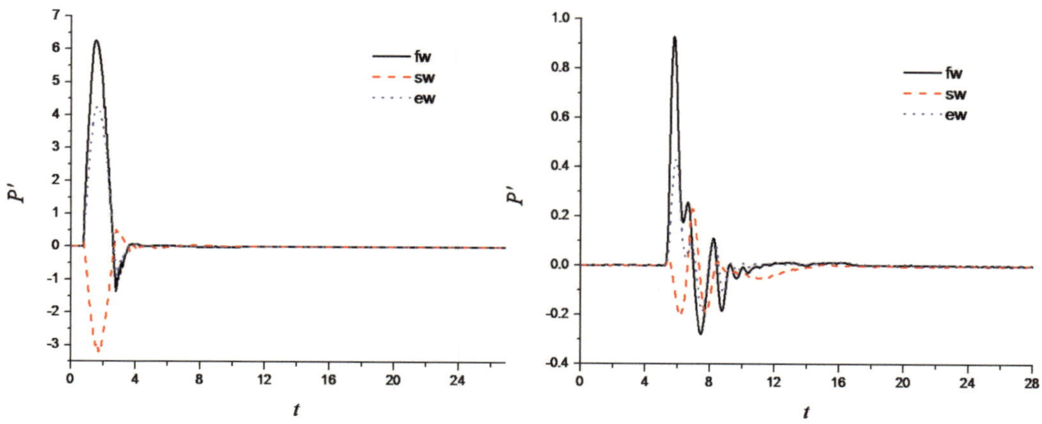

Figure 13. Pressure disturbance in boundary layer under different freestream pulse waves. (a) s=0.92619, (b) s=5.90498.

effects of different types of freestream pulse waves are significant on the distribution of modes in boundary layer and the evolution of disturbance modes along with the streamwise. From the figure, it is obtained that, s = 0.92619, the Fourier frequency spectral curve of pressure disturbance in boundary layer under different freestream pulse waves shows similar trend. Namely, the max amplitude of the disturbance modes in boundary layer are distributed near f = 0.25, and four main disturbance clusters (near f = 0.25, 1.0, 1.5 and 2.0) exist in the hypersonic boundary layer both for all cases. With evolution of disturbance waves in boundary layer along with the streamwise, at s = 3.83551, four main disturbance clusters still exist in the hypersonic boundary layer both for fast acoustic wave and entropy wave. However, the distribution of the four main disturbance clusters changes sharply, which indicates there is movement of main disturbance clusters in the boundary layer. It also can be seen that, at s = 3.83551, there is only one main disturbance cluster (near f = 0.5) in the hypersonic boundary layer for slow acoustic wave, and the other main modes have a remarkable decline. It is obtained that, at s = 8.02802, the number of main disturbance mode clusters in the boundary layer both for fast acoustic wave and entropy wave decreases

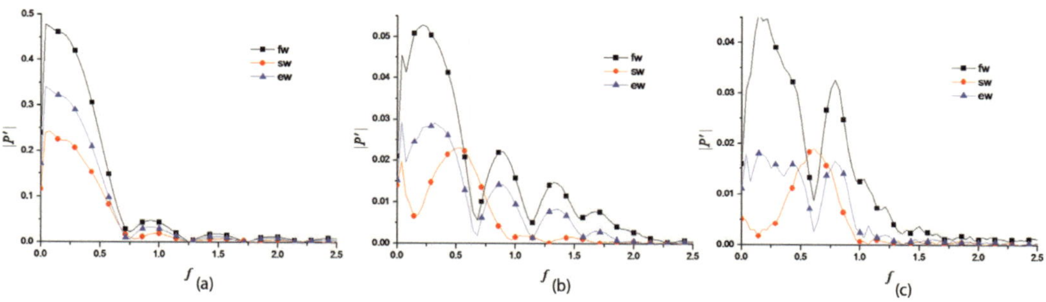

Figure 14. Fourier frequency spectral analysis of pressure disturbance in boundary layer under different freestream pulse waves. (a) s = 0.92619, (b) s = 3.83551, (c) s = 8.02802.

sharply. In general, the narrowing of frequency band and the decreasing of main disturbance mode clusters exist in the boundary layer both for all cases.

5. Conclusions

In present study, a finite difference DNS method is used to investigate the response of the hypersonic flow field to freestream pulse wave, and the generation and evolution characteristics of the disturbance mode in the hypersonic boundary layer are analyzed. The effects of the pulse wave types on the generation and evolution of the disturbance mode in the hypersonic boundary layer are investigated. The study drew the conclusions as follow:

1. The freestream disturbance waves significantly interact with the bow-shaped shock waves, which greatly changes the shock standoff distance and the distribution of flowfield parameters in the active region. The stronger the intensity of the shock wave is, the more intense the effect of the freestream pulse slow acoustic wave and the bow-shaped shock wave are. The freestream pulse wave's influence on the thermodynamic state in boundary layer is significantly greater than that on the thermodynamic state the outside boundary layer. There are many obvious disturbance characteristic regions in the hypersonic boundary layer under the action of freestream pulse disturbances. There is complex interference in the boundary layer under the action of the pulse wave.

2. In the nose area, the main disturbance modes in the boundary layer are distributed near the fundamental frequency, and the disturbance mode component of other frequencies is relatively small. In the nose area, the disturbance mode component ratio near the fundamental frequency decreases rapidly, the high frequency disturbance harmonic mode like the second order harmonic frequency and above significantly increases. With the evolution of the disturbance waves in the boundary layer from the upstream to the downstream, the main disturbance modes in the boundary layer are transformed from the dominant state of the fundamental frequency to the collective leadership state of the second order harmonic frequency and the third order harmonic frequency. The intensity of the bow shock in the hypersonic flowfield has significant effects on both the fundamental mode and the harmonic modes in each order. From the nose area to the non-nose area, the effect of the shock wave intensity on the harmonic frequency modes in the boundary layer is less than the recompression effect, and the effect of the shock wave intensity on the fundamental mode in the boundary layer is larger than the recompression effect.

3. The interaction between the strong shear structure of boundary layer and different types of freestream pulse waves have different mechanisms of action. The strong shear structure of boundary layer under the action of different types of freestream pulse waves reveals different stability characteristics. The effects of different types of freestream pulse waves are significant on the distribution of modes in boundary layer and the evolution of disturbance modes along with the streamwise. The narrowing of frequency band and the decreasing of main disturbance mode clusters exist in the boundary layer both for freestream pulse fast acoustic wave, slow acoustic wave and entropy wave.

Author details

Xiaojun Tang[1]*, Juan Yu[2], Tianli Hui[1], Fenglong Yang[1] and Wentao Yu[1]

*Address all correspondence to: tangxiaojun2214@163.com

1 Beijing Spacecrafts, China Academy of Space Technology, Beijing, China

2 China Center for Information Industry Development, Beijing, China

References

[1] Pierce HB, Manning JC. Experimental Investigation of Blast Loading on an Airfoil in Mach Number of 0.7 Airflow with Initial Angle-of-Attack Change of 20. NASA Resport, 1963, NASA-TN-D-2876.

[2] Aso S, Hayashi K, Mizoguchi M. A study on aerodynamic heating reduction due to opposing jet in hypersonic flow. AIAA Paper. 2002;**2002**:646

[3] Hayashi K, Aso S, Tani Y. Numerical study on aerodynamic heating reduction by opposing jet. Memoirs of the Faculty of Engineering. 2006;**66**(1):39-54

[4] Ma YB, Zhong XL. Boundary-layer receptivity of Mach 7.99 Flow over a blunt cone to free-stream acoustic waves. Journal of Fluid Mechanics. 2006;**556**:55-103

[5] Boiko AV. Receptivity of a flat plate boundary layer to a free stream axial vortex. European Journal of Mechanics - B/Fluids. 2002;**21**:325-340

[6] Zhong X. Leading-edge receptivity to free stream disturbance waves for hypersonic flow over a parabola. Journal of Fluid Mechanics. 2001;**441**:315-367

[7] Fedorov AV, Khokhlov AP. Receptivity of hypersonic boundary layer to wall disturbances. Theoretical and Computational Fluid Dynamics. 2002;**15**:231-254

[8] Tumin A, Wang X, Zhong X. Numerical simulation and theoretical analysis of perturbations in hypersonic boundary layers. AIAA Journal. 2011;**49**(3):463-471

[9] Wang X, Zhong X, Ma Y. Response of a hypersonic boundary layer to wall blowing–suction. AIAA Journal. 2011;**49**(7):1336-1353

[10] Egorov IV, Fedorov AV, Sudakov VG. Direct numerical simulation of disturbances generated by periodic suction-blowing in a hypersonic boundary layer. Theoretical and Computational Fluid Dynamics. 2006;**20**(1):41-54

[11] Kovasznay LSG. Turbulence in supersonic flow. Journal of Aeronautical Sciences. 1953;**20**: 657-682

[12] P. Balakumar. Receptivity of supersonic boundary layers due to acoustic disturbances over blunt cones. 37th AIAA Fluid Dynamics Conference and Exhibit; American Institute of Aeronautics and Astronautics, Miami, Florida; Reston, VA: 2007

[13] Kara K, Balakumar P, Kandil OA. Effects of nose bluntness on hypersonic boundary-layer receptivity and stability over cones. AIAA Journal. 2011;**49**(12):593-2606

[14] Xin-Liang LI, De-Xun FU, Yan-Wen MA, et al. Acoustic calculation for supersonic turbulent boundary layer flow. Chinese Physical Society. 2009;**26**(9):181-184

[15] Ma Y, Zhong X. Receptivity of a supersonic boundary layer over a flat plate, part 2: Receptivity to Freestream sound. Journal of Fluid Mechanics. 2003;**488**(6):79-121

[16] Ma Y, Zhong X. Receptivity of a supersonic boundary layer over a flat plate, part 3: Effects of different types of free-stream disturbances. Journal of Fluid Mechanics. 2005;**532**(6):63-109

[17] Steger J, Warming RF. Flux vector splitting of the in-viscid gasdynamic equations with application to finite difference methods. Journal of Computational Physics. 1981;**40**:263-293

[18] Liu XD, Osher S, Chan T. Weighted essentially non-oscillatory schemes. Journal of Computational Physics. 1994;**115**:200-212

[19] Zhang Y, Fu D, Ma Y, Li X. Receptivity to free-stream disturbance waves for hypersonic flow over a blunt cone. Science in China Series G: Physics, Mechanics & Astronomy. 2008;**51**(11):1682-1690

[20] Shu CW. Essentially Non-Oscillatory and Weighted Essentially Non-Oscillatory Schemes for Hyperbolic Conservation Laws. NASA/CR-97-206253

[21] Tang X, Zhu X, Hui T, et al. Receptivity characteristics of a hypersonic boundary layer under freestream slow acoustic wave with different amplitudes. European Physical Journal Applied Physics. 2017;**79**:31101-1-31101-15

[22] Tang X, Lv H, Meng X, Wang Z, Lv Q. Stability characteristic of hypersonic flow over a blunt wedge under freestream pulse wave. Central. European Journal of Physics. 2014;**12**(1): 17-31

[23] Wang Z, Tang X, Lv H, Shi J. Temporal and spatial evolution characteristics of disturbance wave in a hypersonic boundary layer due to single-frequency entropy disturbance. The Scientific World Journal. 2014;**39**(2):359-371. DOI: 10.1155/2014/517242

[24] Xian L, Li XL, Fu DX, Ma YW. Effects of wall temperature on boundary layer stability over a blunt cone at Mach 7.99. Computers & Fluids. 2010;**39**:359-371

[25] Saric WS, Reed HL, Kerschen EJ. Boundary-layer receptivity to freestream disturbances. Annual Review of Fluid Mechanics. 2002;**34**:291-319

Chapter 3

Numerical Modeling of Hypersonic Aerodynamics and Heat Transfer Problems of the Martian Descent Modules

Yuriy D. Shevelev

Additional information is available at the end of the chapter

http://dx.doi.org/10.5772/intechopen.71666

Abstract

Computational fluid dynamics (CFD) is the important tool to analyze physics of fluids. Hypersonic flows over real space configurations represent a substantial problem from the point of view of the development of new and more effective mathematical models, numerical methods, and the use of computer systems. Governing equations for multi-component, multi-temperature, chemically reacting non-equilibrium radiant mixtures are the mathematical foundation for the study of vehicles entering in Martian atmosphere. Based on the kinetic equations for the distribution functions, an efficient three-temperature model suitable for Mars re-entry applications derived and used for the simulations of a non-equilibrium flow in a viscous shock layer near a space vehicle. The closed self-consistent description of a flow in terms of densities of species, macroscopic velocity, gas temperature, and three vibration temperatures are proposed. The transport properties in dissociating CO_2 flows have been evaluated. The proposed model takes into account real structure of polyatomic CO_2 molecules, non-equilibrium CO_2 vibration excitation, different rates of various energy transitions, and chemical reactions. Numerical investigations of a flow past a frontal part of Mars Sample Return Orbiter (MSRO) and MARS EXPRESS vehicles descending in an atmosphere of Mars are presented. The radiation processes taking into account of non-equilibrium character are considered.

Keywords: mechanical and aerospace engineering, aerothermophysics, thermo-chemical models, convective and radiation heat transfer

1. Introduction

A problem of non-equilibrium thermochemical processes in carbon dioxide (CO_2) mixture flows has been studied in a great number of experimental and theoretical works owing to needs of Mars planet investigations. These works started in the 1960s. Further studies were stimulated by development of Pathfinder and other Martian entry vehicles [1–6].

There are two ways of derivation of the gas dynamics equations for a multi-component mixture: kinetic and phenomenological. The first approach is based on the kinetic theory of gases. The zero-order and the first-order distribution functions are known and expressed by means of different gradients, the diffusive driving forces, etc. [7, 8]. If potentials of interaction between particles are defined, then the equations in terms of species densities, macroscopic velocity, and gas temperature can be carried out by methods of the kinetic theory of gases [9–12]. The transport theory of polyatomic gas mixtures taking into of account internal molecular structure, different rates of vibration, and non-harmonic transitions has been developed for a five component $CO_2/O_2/CO/O/C$ mixture taking vibration excitation of molecules into account. The algorithm for transport coefficients calculation has been realized as a program module.

The phenomenological approach is based on application of the basic theorems of mechanics and laws of thermodynamics for macroscopic volume of the continuous media and leads to unclosed system of the equations. For closure of this system, the additional definition of mass, momentum, and energy fluxes is necessary. Unknown values of transfer coefficients within the framework of the phenomenological approach can be defined from approximation either from experiment. The kinetic and phenomenological approaches give the same result as structure of gas dynamic equations system.

The governing equations with rigorous kinetic schemes for transport coefficients are solved numerically for a flow in a viscous shock layer near the blunt body imitating the form of the spacecraft Mars Sample Return Orbiter (MSRO) and MARS EXPRESS vehicles descending in an atmosphere of Mars for the conditions typical for the re-entering regime [9–12]. The form of considered vehicles represents spherically blunted cone with aт angle 120 and radius $R = 1.0$ m (MSRO) and $R = 0.38$ m (MARS EXPRESS) joined with the cylinder (**Figure 1**).

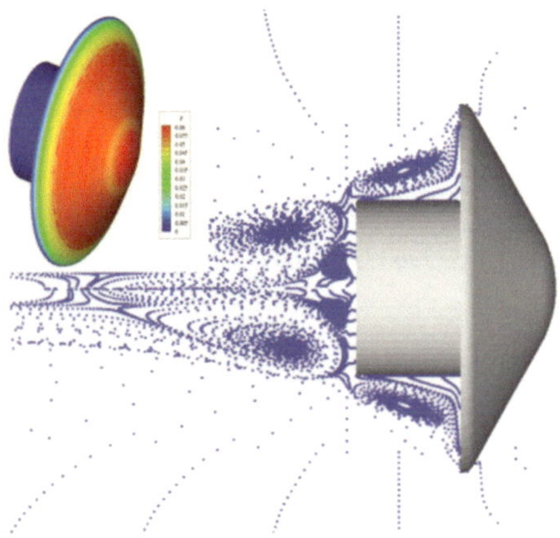

Figure 1. Martian vehicle.

Calculations of the convective heat flux and the non-equilibrium radiation were carried out of the MSRO vehicle entering into the Martian atmosphere [7–19].

The part of results have been obtained in the framework of the INTAS Project 03-51-5204 "Numerical, Analytical and Experimental Investigation of Convective and Radiative Heating of Martian Descent Module". This Project was elaborated jointly between institutes-participants from Russian side and European Collaborators [5].

2. Governing equations of hypersonic non-equilibrium polyatomic gas flows

The Martian atmosphere is composed mostly of carbon dioxide (96%), nitrogen (1.9%), argon (1.9%), and others. Small admixtures of nitrogen (N_2) and argon (Ar) in the Mars atmosphere do not play a significant role in the process of heat transfer to descent vehicles (at least, at moderate velocities of flight till the convective heat transfer prevails). It is possible to restrict ourselves by consideration of model atmosphere as the pure carbon dioxide. The pressure on the planet surface is taken equal to 6.0 mbar. It is 0.6% of Earth's mean sea level pressure. The atmosphere is quite dusty.

The conditions of a flow corresponding to the last stage of flight of space vehicles in an atmosphere of Mars ($V_\infty \leq 6$ km/s, $\rho_\infty > 10^{-5}$ kg/m^3, $H < 60$ km) were studied. Determining process at such velocities is a process of dissociation. Up to 75% of full gas flow energy can be spent on it.

The region where non-equilibrium physical and chemical processes realized is a significant part from all considered regions (**Figure 2**). Velocity of physical and chemical processes, as a rule, grows together with density of gas. For considered flow conditions, the degree of gas ionization is small and does not bring the appreciable contribution to internal gas energy. The translational degrees freedom becomes equilibrium on distances of several free path lengths of molecules behind front of a shock wave for considered altitude. The distribution of rotation energy also is established slightly later. Therefore, it is usually supposed that translational and vibration degrees of freedom of particles are in equilibrium. At high temperatures that observed in a shock layer, the characteristic times of a vibration energy relaxation of molecules and characteristic times of dissociation become one order. Thus the account of non-equilibrium excitation of vibration degrees of freedom of carbon dioxide molecules is necessary. The region of relaxation behind the bow shock wave has a specific structure that consists of sequential relaxation zones. The flow in shock and boundary layers are being especially non-equilibrium. It causes the energy redistribution of the internal energy.

Since the Martian atmosphere is strongly rarefied, non-equilibrium processes affect heat transfer at the more significant part of the descent trajectory. The most thermal-loaded part of the typical descent trajectory is the region of frozen chemical reactions and equilibrium-excited vibration degrees of freedom.

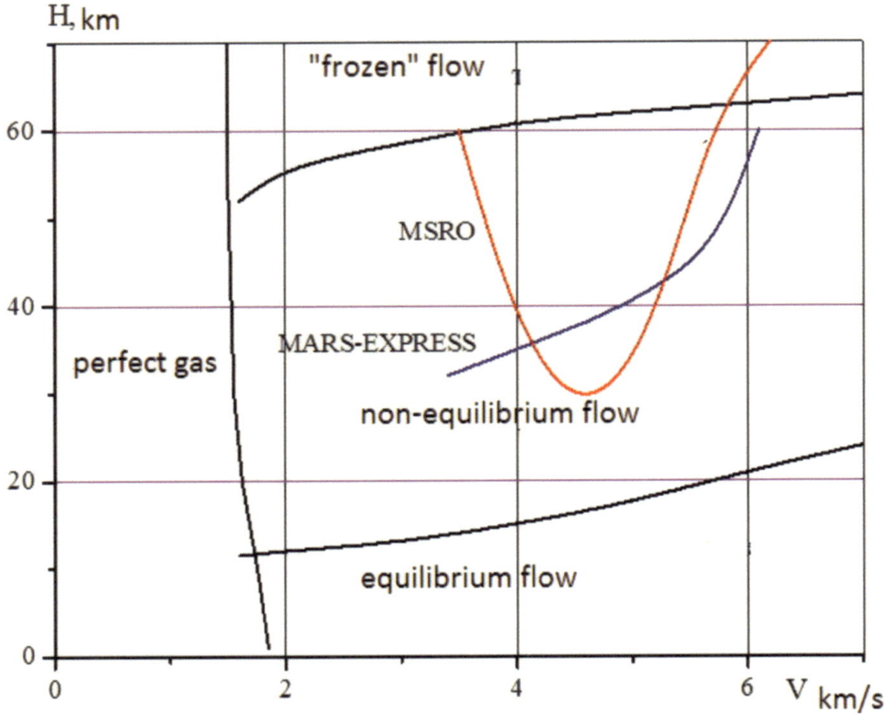

Figure 2. (a) The region of different flow regimes of flow for MSRO vehicle (red line) and MARS EXPRESS vehicle (blue line); (b) Mars descent vehicles.

The thermo-chemical model of the processes occurring in the shock layer includes the chemical reactions, dissociation and recombination of CO_2 molecules, dissociation and recombination of diatomic molecules, exchange reactions, processes of vibration energy exchange between various levels of molecules, influence of the vibration relaxation on the chemical reactions and vice versa (CVDV-processes), processes of excitation and deactivation of the electronic states of molecules, and spontaneous radiation processes for excited particles.

We consider a high-temperature flow of the mixture taking into account vibration excitation and the following exchange reactions, dissociation, and recombination:

$$CO_2 + M \leftrightarrow CO + O + M, \circledcirc CO + M \leftrightarrow C + O + M, \quad (1)$$

$$O_2 + M \leftrightarrow O + O + M, \circledcirc CO_2 + O \leftrightarrow CO + O_2, \quad (2)$$

$$CO + CO \leftrightarrow CO_2 + C, \circledcirc CO + O \leftrightarrow O_2 + C, \quad (3)$$

where M is a molecule or an atom [10].

As known molecule of CO_2 have three vibration modes with different characteristic temperatures: symmetric, deformation (twice degenerate), and asymmetric. According to fast exchange of vibration energy between the different modes, it is assumed that molecules reach the Boltzmann distribution with a uniform temperature.

We consider the conditions typical for a high-temperature shock layer, while the translational and rotational relaxation are supposed to proceed fast as well as intra-mode VV-transitions in CO_2, O_2, CO and inter-mode VV-exchange between CO_2 symmetric and bending modes. All other vibration energy transitions, dissociation, recombination, and exchange reactions are considered to be slower with relaxation times comparable with the mean time of the gas dynamic parameters variation. Such a relation between the characteristic times makes it possible to introduce vibration temperatures for the coupled (symmetric-bending) and asymmetric CO_2 modes.

The existing experimental and theoretical data on relaxation times of different processes in mixtures containing CO_2 molecules show that in a wide range of conditions the following relations are valid:

$$\tau_{tr} < \tau_{rot} < \tau_{VV_m} \sim \tau_{VV'_{1-2}} << \tau_{VT_2} \sim \tau_{VV'_{1-2-3}} < \tau_{VT_3} << \tau_r \sim \theta, \quad m = 1,\ 2,\ 3 \tag{4}$$

Here τ_{tr}, τ_{rot} are the characteristic times of translational and rotation relaxation; τ_{VV_m} are the times of intra-mode VV exchanges; τ_{VT_2}, τ_{VT_3} are the times of VT transitions; $\tau_{VV'_{k-m}}$ correspond to inter-mode transitions; τ_r is the characteristic time of chemical reactions; and θ is the mean time of gas dynamic parameters change.

On the basis of the kinetic theory principles, the closed self-consistent three-temperature description of a flow in terms of densities of species, macroscopic velocity, gas temperature, and three vibration temperatures are obtained [7, 8]. The set of governing equations contains the conservation equations of mass, momentum, and total energy coupled to the equations of non-equilibrium three-temperature chemical kinetics as well as the relaxation equations for vibration temperatures.

Under condition (5) the set of equations are obtained in the following form:

$$\frac{d\rho}{dt} + \rho\ \nabla \cdot \mathbf{v}\ = 0 \tag{5}$$

$$\rho\frac{d\mathbf{v}}{dt} + \nabla \cdot \mathbf{P} = 0 \tag{6}$$

$$\rho\frac{de}{dt} + \nabla \cdot \mathbf{q} + \mathbf{P} : \nabla\mathbf{v} = 0 \tag{7}$$

The equations of non-equilibrium chemical kinetics written in the following form

$$\frac{dn_i}{dt} + n_i\nabla \cdot \mathbf{v} + \nabla \cdot (n_i\mathbf{V}_i) = R_i^r\ \ (i = CO_2, CO, O_2, O, C) \tag{8}$$

Here n_i is the number density of species i (1—CO_2, 2—CO, 3—O_2, 4—C, 5—O). Values $\rho_i = m_i n_i$ is the species density, m_i and n_i are the species mass and number density, $\rho = \sum \rho_i$ is the mixture density; $\mathbf{v}$ is the macroscopic gas velocity, e is the mixture total energy per unit mass; R_i^t, R_{12}, R_3 are the production terms due to dissociation, recombination, exchange reactions and slow processes of CO_2 vibration relaxation; $\mathbf{V}_i$ is the diffusion velocity; $\mathbf{P}$ is the pressure tensor; and $\mathbf{q}$ is the heat flux.

We consider the condition that corresponds to rapid translational and rotational relaxation, VV_m is the vibration energy exchanges within modes and VV'_{12} is the exchange between symmetric and bending CO_2 modes. In this case the vibration CO_2 distributions depend on the vibration temperatures T_{12} of the combined (symmetric + bending) mode, and T_3 of the asymmetric mode. The vibration distributions of CO and O_2 are supposed to be close to the thermal equilibrium; vibration spectra are simulated by the harmonic oscillator model. Values $E_{vibr1}(T_{12},T_3) = E_{12}(T_{12}) + E_3(T_3)$, $E_{vibr2}(T)$, $E_{vibr3}(T)$ are the specific vibration energies of molecular species CO_2, CO, and O_2, respectively; thus $E_{12}(T_{12})$, $E_3(T_3)$ are the specific vibration energies of non-equilibrium CO_2 modes.

The equations of non-equilibrium vibration kinetics written in the following form

$$\rho_1 \frac{dE_{12}}{dt} + \nabla \cdot \mathbf{q}_{12} = R_{12} - m_1 E_{12} R_1^r + E_{12} \nabla \cdot (\rho_1 \mathbf{V}_1) \tag{9}$$

$$\rho_1 \frac{dE_3}{dt} + \nabla \cdot \mathbf{q}_3 = R_3 - m_1 E_3 R_1^r + E_3 \nabla \cdot (\rho_1 \mathbf{V}_1) \tag{10}$$

x_i is the molar fraction of species i (1—CO_2, 2—CO, 3—O_2, 4—C, 5—O). Values $\mathbf{q}_{12}$, $\mathbf{q}_3$ are the fluxes of vibration energy in the combined and asymmetric modes, respectively.

The vibration relaxation of molecules time of CO_2 molecules is calculated under the usual formulas by approximation of theoretical and experimental data for particles of different types.

3. Transport and source terms

The transport theory of polyatomic gas mixtures taking into account internal molecular structure, different rates of vibration transitions, and unharmonicity has been developed for a five component $CO_2/O_2/CO/O/C$ mixture taking into account vibration excitation of diatomic molecules.

The transport properties in the viscous gas approximation are determined by the first-order distribution functions. The zero-order and the first-order distribution functions are known and express by means different gradients, the diffusive driving forces, etc. [7, 8].

Pressure tensor, diffusion velocity, heat flux, and vibration energy fluxes are expressed in terms of macroscopic parameters gradients and transport coefficients.

The pressure tensor is obtained in the form:

$$\mathbf{P} = \left(p - p_{rel}\right) \mathbf{I} - 2\eta \ \mathbf{S} - \zeta \nabla \cdot \mathbf{v} \mathbf{I}. \tag{11}$$

Here $\mathbf{S}$ is the strain rate tensor, $\mathbf{I}$ is the unit tensor, η is the shear viscosity coefficients, and ζ and p_{rel} are the bulk viscosity coefficient and relaxation pressure appearing in the diagonal elements of the pressure tensor due to rapid inelastic non-resonant processes. The Navier-Stokes equation derivation leads to the definition of two coefficients of viscosity: shear coefficient of viscosity and coefficient of bulk viscosity.

Transport coefficients (bulk and shear viscosity, heat conductivity, diffusion, pressure- and thermo-diffusion of multi-component gas mixture) are calculated according to the basic kinetic theory. The algorithms for thermal conductivity, vibration thermal conductivity, diffusion, thermal diffusion, shear, and bulk viscosity coefficients computation are developed.

These transport coefficients are defined by functions B_{cij}, F_{cij}, G_{cij}:

$$\eta = kT[B,B]/10, \quad \zeta = kT[F,F], p_{rel} = kT[F,G] \tag{12}$$

Here $[A;B]$ are the bracket integrals depending on the cross sections of rapid processes (see for instance [7, 8] for definition).

The values of diffusion fluxes can be written down through thermodynamic forces (external mass forces it is neglected):

$$\mathbf{J}_i = -\rho_i \sum_j D_{ij}^* \mathbf{d}_j - \rho_i D_{T_i} \nabla \ln T \tag{13}$$

Here D_{ij}^* is the multi-component coefficients of diffusion, D_{T_i} is the coefficients of thermo-diffusion, $\mathbf{d}_j$ is the diffusion driving forces:

$$\mathbf{d}_j = \nabla\left(\frac{n_j}{n}\right) + \left(\frac{n_j}{n} - c_j\right)\nabla \ln p. \tag{14}$$

Here n_j is the number of jth mole components, n is the common mole's number. Let us take the difference between D_{ij}^* and D_{ij}, where D_{ij} is the binary diffusion coefficients.

The basic way of finding the independent transfer coefficients D_{ij}^*, D_{T_i} are connected to the kinetic theory methods. In order to define the multi-components diffusion coefficients, it is necessary to solve the system of the linear algebraic equations that in case of gas mixture have the next form

$$D_{ij} = \left[\mathbf{D}^{\mathbf{i}}, \mathbf{D}^{\mathbf{i}}\right]/3n, \quad D_{Ti} = \left[\mathbf{D}^{i}, \mathbf{A}^{\mathbf{i}}\right]/3n. \tag{15}$$

Here $\left[\mathbf{D}^{\mathbf{i}}, \mathbf{D}^{\mathbf{i}}\right]$, $\left[\mathbf{D}^{i}, \mathbf{A}^{\mathbf{i}}\right]$ are the bracket integrals depending on the cross sections of rapid processes. Functions $\mathbf{A}_{cij}$, $\mathbf{A}_{cij}^{(12)}$, $\mathbf{A}_{cij}^{(3)}$, $\mathbf{A}_{cij,}^{(c,1)}\mathbf{B}_{cij}$, F_{cij}, F_{cij} are found from the linear integral equations for the first-order correction to the distribution function. The relaxation pressure determine by the slow non-equilibrium processes. This quantity is basically supposed to be small compared to p, and by value p_{rel} is neglected.

The total heat transfer of a multi-component mixture is defined by effects of heat conductivity of various kinds of energy and diffusion. The heat flux is given by the formula

$$\mathbf{q} = -\lambda'\nabla T - \lambda_{12}\nabla T_{12} - \lambda_3\nabla T_3 - p\sum_{i=1}^{5} D_{T_i}\mathbf{d}_i + \sum_{i=1}^{5}\rho_i h_i \mathbf{V}_i \tag{16}$$

Here p is the pressure, h_i is the specific enthalpy of species i, $\lambda' = \lambda_{tr} + \lambda_r + \lambda_v$ is the thermal conductivity coefficient of all degrees of freedom which deviate weakly from local thermal equilibrium. They include the translational and rotational modes as well as CO and O_2 vibration degrees of freedom. Thus, the coefficient $\lambda_v = \lambda_{v,CO} + \lambda_{r,O2}$. Coefficients λ_{12}, λ_3 correspond to the thermal conductivity of strongly non-equilibrium modes: combined (symmetric + bending) and asymmetric ones.

The fluxes of vibration energy in the combined and asymmetric CO_2 modes in the harmonic oscillator approach depend only on the gradient of corresponding vibration temperature:

$$\mathbf{q}_{12} = -\lambda_{12}\nabla T_{12}, \quad \mathbf{q}_3 = -\lambda_3\nabla T_3 \tag{17}$$

The thermal conductivity coefficients are expressed in terms of bracket integrals

$$\lambda = k[\mathbf{A},\mathbf{A}]/3, \quad \lambda_{12} = k\left[\mathbf{A}^{(12)},\mathbf{A}^{(12)}\right]/3, \lambda_3 = k\left[\mathbf{A}^{(3)},\mathbf{A}^{(3)}\right]/3 \tag{18}$$

The algorithms of transport coefficients calculation are similar for various multi-temperature models and consist of the following steps:

1. Unknown functions $A_{cij}, A_{cij}^{(12)}, A_{cij}^{(3)}, A_{cij,}^{(c,1)} B_{cij}, F_{cij}, F_{cij}$ are expanded in the series of Sonine and Waldmann-Trubenbacher polynomials; the trial functions are introduced accordingly to the right hand sides of integral equations for the first-order correction to the distribution function.

2. Transport coefficients are expressed in terms of expansion coefficients.

3. Integral equations are reduced to the linear systems of algebraic equations involving bracket integrals as coefficients.

4. Bracket integrals are simplified on the basis of some assumptions about cross sections of rapid processes; finally they are expressed in terms of the standard $\Omega_{cd}^{(l,r)}$ is the integrals and relaxation times which can be measured experimentally. The $\Omega_{cd}^{(l,r)}$ is the integrals are calculated for particular models of inter-molecular interaction potentials. In the present study, the Lennard-Jones potential is used for low and moderate temperatures whereas in the high-temperature interval the repulsive potential is applied.

5. Transport coefficients are found as solutions of transport linear systems using some numerical algorithms (for instance, the Gauss method or new iterative procedures).

For example, let us consider the bulk viscosity coefficient. In Navier-Stokes equations, the terms involving bulk viscosity multiplied by divergence of velocity and can play a significant role in flow fields with substantial dilatation. The bulk viscosity coefficient was discussed

http://dx.doi.org/10.5772/intechopen.71666

theoretically in the literature [18] for weak and strong non-equilibrium conditions but up to now it was not evaluated in real gas flows.

The experiments and kinetic theory show that bulk viscosity can significantly influence shock wave structure for polyatomic molecules. In polyatomic gases of the deviation from local equilibrium effects itself as bulk viscosity. From the Chapman-Enskog's theory, it can be proved that for any perfect monoatomic gas, the coefficient of bulk viscosity is equal to zero. Bulk viscosity results from contributions from the several internal degrees of freedom of the gas.

The bulk viscosity coefficient is defined by rotational energy transitions of all molecular species and VT vibration energy transfer in CO and O_2 molecules and can be written as follows

$$\zeta = \zeta_r + \zeta_{v,CO} + \zeta_{v,O_2} \tag{19}$$

For harmonic oscillators, rapid inelastic VV and VV'_{1-2} exchanges occur to be resonant and therefore do not give contribution to the coefficient ζ.

The bulk viscosity coefficient can be obtained in the form [18]:

$$\zeta = -kT\left(x_1 f_{1,10} + x_2 f_{2,10} + x_3 f_{1,10} + x_4 f_{4,1} + x_5 f_{5,1}\right), \tag{20}$$

where k is the Boltzmann constant, x_i is the molar fraction of species i (1—CO_2, 2—CO, 3—O_2, 4—C, 5—O).

Coefficients $f_{i,mn}$ are the solutions of the system

$$\mathbf{B} \times \mathbf{f} = \mathbf{s}. \tag{21}$$

Here matrix **B** is composed of the bracket integrals $\beta^{ij}_{mm'nn'}$:

$$\mathbf{B} = \begin{pmatrix} x_1 c_{tr} & x_2 c_{tr} & x_3 c_{tr} & x_4 c_{tr} & x_5 c_{tr} & x_1 c_{rot,1} & x_2 c_{\text{int},2} & x_3 c_{\text{int},3} \\ \beta^{2-1}_{11} & \beta^{2-2}_{11} & \beta^{2-3}_{11} & \beta^{2-4}_{11} & \beta^{2-5}_{11} & \beta^{2-1}_{1001} & \beta^{2-2}_{1001} & \beta^{2-3}_{1001} \\ \beta^{3-1}_{11} & \beta^{3-2}_{11} & \beta^{3-3}_{11} & \beta^{3-4}_{11} & \beta^{3-5}_{11} & \beta^{3-1}_{1001} & \beta^{3-2}_{1001} & \beta^{3-3}_{1001} \\ \beta^{4-1}_{11} & \beta^{4-2}_{11} & \beta^{4-3}_{11} & \beta^{4-4}_{11} & \beta^{4-5}_{11} & \beta^{4-1}_{1001} & \beta^{4-2}_{1001} & \beta^{4-3}_{1001} \\ \beta^{5-1}_{11} & \beta^{5-2}_{11} & \beta^{5-3}_{11} & \beta^{5-4}_{11} & \beta^{5-5}_{11} & \beta^{5-1}_{1001} & \beta^{5-2}_{1001} & \beta^{5-3}_{1001} \\ \beta^{1-1}_{0110} & \beta^{1-2}_{0110} & \beta^{1-3}_{0110} & \beta^{1-4}_{0110} & \beta^{1-5}_{0110} & \beta^{1-1}_{0011} & 0 & 0 \\ \beta^{2-1}_{0110} & \beta^{2-2}_{0110} & \beta^{2-3}_{0110} & \beta^{2-4}_{0110} & \beta^{2-5}_{0110} & 0 & \beta^{2-2}_{0011} & 0 \\ \beta^{3-1}_{0110} & \beta^{3-2}_{0110} & \beta^{3-3}_{0110} & \beta^{3-4}_{0110} & \beta^{3-5}_{0110} & 0 & 0 & \beta^{3-3}_{0011} \end{pmatrix} \mathbf{f} = \begin{pmatrix} f_{1,10} \\ f_{2,10} \\ f_{3,10} \\ f_{4,1} \\ f_{5,1} \\ f_{1,01} \\ f_{2,01} \\ f_{3,01} \end{pmatrix} \mathbf{s} = \begin{pmatrix} 0 \\ -x_2(c_u - c_{tr})/c_u \\ -x_3(c_u - c_{tr})/c_u \\ -x_4(c_u - c_{tr})/c_u \\ -x_5(c_u - c_{tr})/c_u \\ x_1 c_{rot,1}/c_u \\ x_2 c_{\text{int},2}/c_u \\ x_3 c_{\text{int},3}/c_u \end{pmatrix} \tag{22}$$

Matrix **B,** vectors **f** and **s** are given above. Here c_{tr}, c_{rot}, c_{int} are specific heats of translation, rotational, internal degrees of freedom and c_u is the total specific heat. The bracket integrals

$\beta^{ij}_{mm'nn'}$ required for the evaluation of the bulk viscosity coefficient are determined by the energy variation in inelastic processes. The system (22) in this form has a unique solution.

The rates of vibration energy transitions are expressed in terms of corresponding relaxation times. The rate coefficients for non-equilibrium CO_2 dissociation were calculated using the expressions proposed in Ref. [8] as an extension of the Treanor-Marrone's model [20] for three-atomic molecules. For the recombination rate coefficients, the detailed balance principle is used. For the rate coefficients of exchange reactions and dissociation of diatomic molecules, the Arrhenius formulas are applied. The vibration relaxation of molecules time of CO_2 molecules is calculated under the usual formulas by approximation of theoretical and experimental data for particles of different types [10].

4. Results the calculation of transport coefficients

For calculation of transport coefficient of gas mixture, ones need the information how interaction between themselves and each pair of species of gas mixture is going on. It is shown that under considered conditions, the transport coefficients are defined by the collisions with translational and rotational energy changing whereas the reaction rate coefficients depend on the cross sections of slow energy transitions, dissociation, and exchange reactions.

The algorithm for the calculation of transport coefficients has been realized for the 5-temperature model as a program module in a form of Fortran 90 code. The code calls several independent modules: CONSTANT: common constants and variables definition; SPECIFIC HEAT: calculates vibration energy levels, non-equilibrium vibration partition functions, vibration specific heat capacities; OMEGA INTEGRALS: calculates integrals and their ratios using the Lennard-Jones and the Born-Meyer potentials for moderate and high-temperature ranges; BRACKET INTEGRALS: calculates bracket integrals in the transport linear systems; INVERS: solves systems of linear algebraic equations using the Gauss method.

Eqs. (6)–(11) with kinetic schemes for transport coefficients described above are solved numerically for a flow in a viscous shock layer near the blunt body imitating the form of the spacecraft MSRO (Mars Sample Return Orbiter) for the conditions typical for the re-entering regime.

In **Figure 3**, coefficients of shear and volume viscosity along a stagnation line are presented. Calculations are obtained for ideal catalytic wall having the constant temperature T = 1500 K and conditions of a flow of MSRO vehicle $V_\infty = 5223$ m/s, $\rho_\infty = 2.93 \times 10^{-4}$ kg/m^3 (curves T1) and $V_\infty = 5687$ m/s, $\rho_\infty = 3.141 \times 10^{-5}$ kg/m^3 (curves T2). Near surface of a body, the value of volume viscosity about value of shear, and in a shock layer surpasses it approximately in two times. It is established that the mechanism of non-equilibrium excitation of vibration degrees of freedom of molecules CO_2 does not affect on value of volume viscosity. Estimations of influence of volume viscosity on parameters of flow and a heat transfer to a surface of a space vehicle in an atmosphere of Mars are carried out. For the specified parameters of a flow, the

http://dx.doi.org/10.5772/intechopen.71666

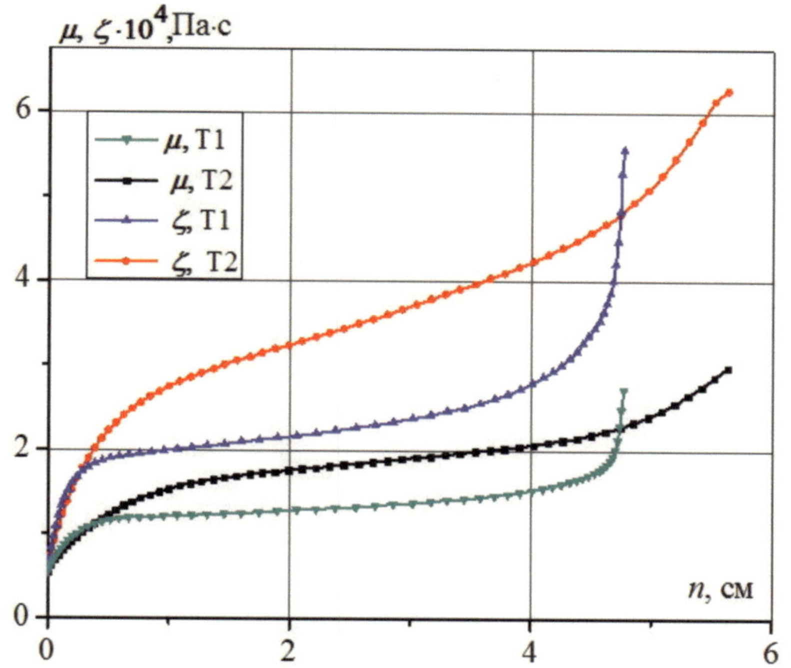

Figure 3. Coefficient of shear μ and volume ζ viscosity along a stagnation line. T1: $V_\infty = 5223$ m/s, $\rho_\infty = 2.93 \times 10^{-4}$ kg/m^3, T2: $V_\infty = 5687$ m/s, $\rho_\infty = 3.141 \times 10^{-5}$ kg/m^3.

influence of account of volume viscosity in the equations of flow leads to increase of a heat flux up to 10%. The similar tendency takes place and for other conditions of a flow.

In **Figure 4**, the similar data are presented for coefficient of heat conductivity.

For conditions of a MSRO vehicle, flow values of multi-component diffusion coefficient D^*_{ij} have been obtained with help of diffusion flux definition through thermodynamic forces [13].

In **Figure 5**, the distribution of self-diffusion coefficients along a stagnation line for the some component of a gas mixture is presented.

The values corresponding to diagonal elements D^*_{ij} $(i \neq j)$ of a matrix diffusion, and in **Figure 6** shows non-diagonal elements D^*_{ij} $(i \neq j)$. It is evident that values of the elements belonging to the main diagonal in the most part of a shock layer surpass values of non-diagonal elements. It is testifies to legitimacy of application of the Fick's law for calculation of diffusion flux. However, near surface of a body and in the field of a shock wave the values of elements (e.g. D^*_{11} and D^*_{1j}, $j \neq 1$) can be same order. It means that in these zones to use Fick's law it is incorrect.

Diffusion flux of everyone components depends on own gradient of concentration components and coefficient of self-diffusion. In **Figure 7** confirmation of this fact are presented and the diffusion velocities for component CO_2 and CO obtained by using the "exact" expression

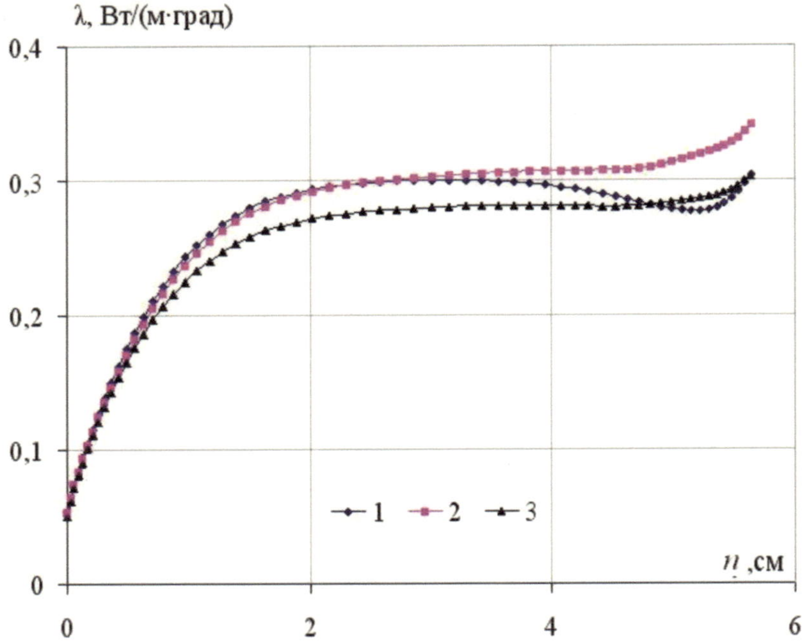

Figure 4. Coefficient of heat conductivity along a stagnation line for MSRO vehicle. 1: model of [8], 2: model of [21, 22], 3: model of [23, 24]. $V_\infty = 5687$ m/s, $\rho_\infty = 3.141 \times 10^{-5}$ kg/m^3.

$$V_i = -\sum_j D^*_{ii} \nabla x_j \tag{23}$$

and relation

$$V_i = -D^*_{ii} \nabla x_i. \tag{24}$$

From the data of **Figure 7**, it is obtained that for considered flow conditions, it is necessary to take into account and non-diagonal elements of diffusion matrix. Influence of thermo-diffusion and pressure diffusion on parameters of a flow was considered also. As pressure across a shock layer is equal practically constant then process of pressure diffusion can be not taken into account. The temperature in a shock layer changes essentially.

The temperature gradients are observed near a body surface and near a shock wave. In **Figure 8**, distribution along a stagnation line of sizes of thermo-diffusion coefficient D_T for separate component of a gas mixture is shown. In the first case, diffusion velocity speed was calculated under the formula

$$V_i = -\sum_j D^*_{ij} \nabla x_j - D_{T_i} \nabla \ln T \tag{25}$$

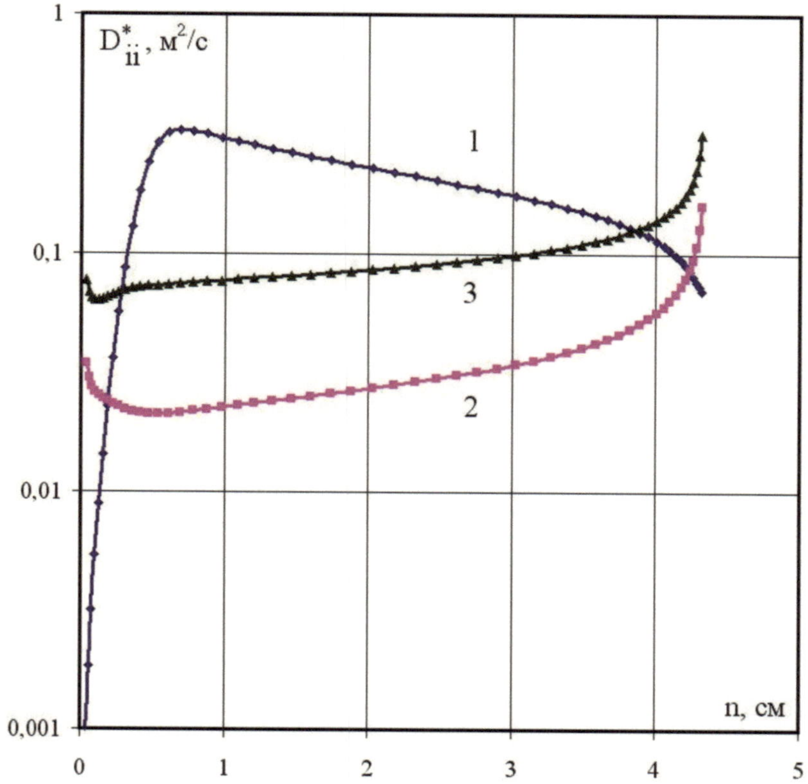

Figure 5. The coefficient of self-diffusion D^*_{ii} along stagnation line. 1: D^*_{11} (CO_2–CO_2), 2: D^*_{22} (CO–CO), 3: D^*_{55} (O–O). $V_\infty = 5223$ m/s, $\rho_\infty = 2.933 \times 10^{-4}$ kg/m^3.

In the second case, without taking into account the second term in the right part of above expression. The data resulted in **Figure 8** confirm that influence of thermo-diffusion effect is small.

In **Figure 9**, the obtained values of diffusion velocities for various component of a mixture with taking in account and without of thermo-diffusion are presented. It is evident that these values basically are much lower than corresponding parameters of mass diffusion. It allows suppose that thermo-diffusion influences are negligible. However for full clearness, it is necessary to take into account change of temperature.

In **Figure 10**, comparison of effective diffusion coefficients D_i for a component of mixture CO_2 and CO is determined in two ways—with the help of binary diffusion coefficients D_{ij} and multi-component coefficients D_{ij} is shown. The data in **Figure 10** are presented along a stagnation line across a shock layer for conditions of a flow of the vehicle: V_∞ = 5292 m/s, $\rho_\infty = 2.5 \times 10^{-4}$ kg/m^3 in a case of ideal catalytic surfaces. It is shown that the effective diffusion coefficient determined with the help of two methods is very close.

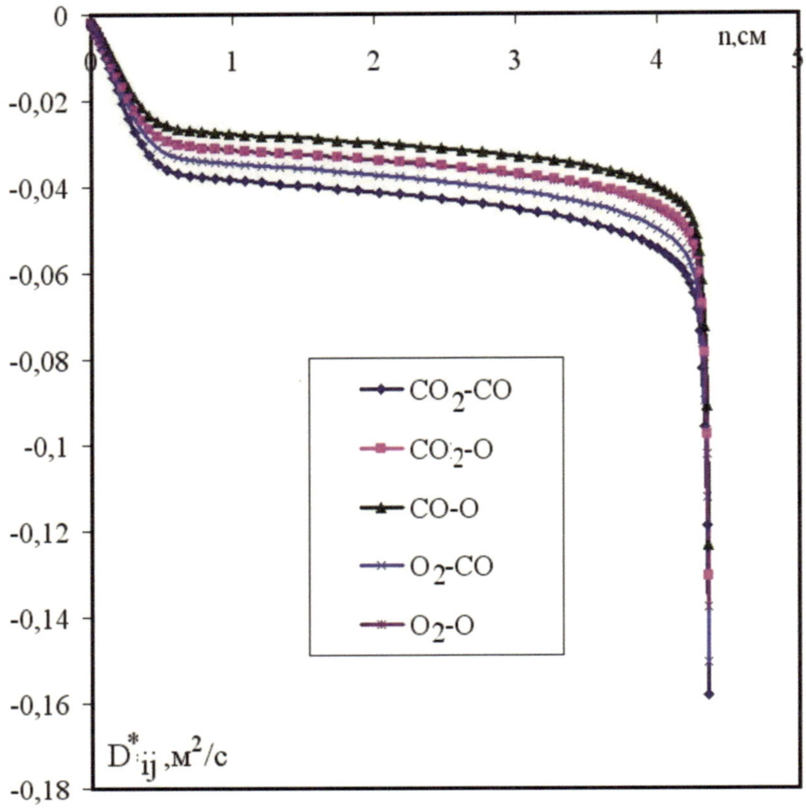

Figure 6. Coefficient self-diffusion D^*_{ij} $(i \neq j)$ along stagnation line, $V_\infty = 5223$ m/s, $\rho_\infty = 2.933 \times 10^{-4}$ kg/m^3.

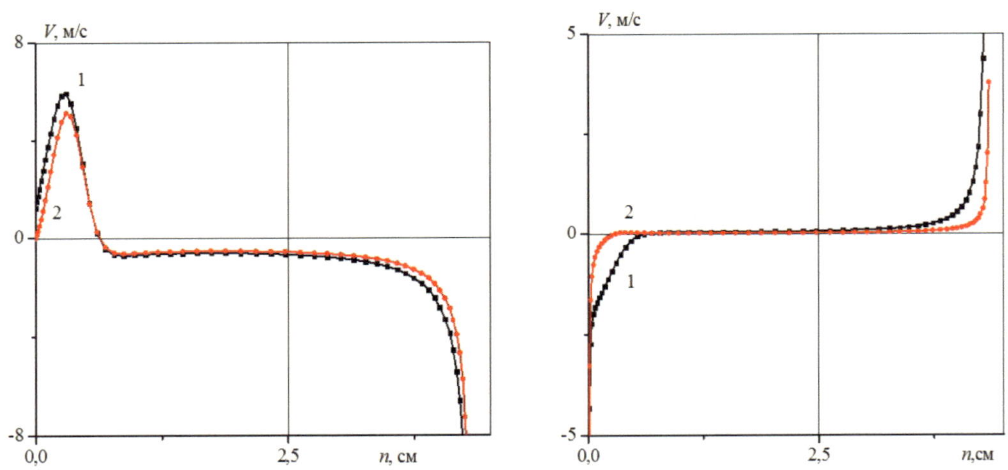

Figure 7. Diffusion velocity along the stagnation line: 1: results with taking into account of all diffusion coefficients (formula (24)); 2: results with taking into account coefficients self-diffusion (formula (25)); (a) red line—CO_2 component; (b) black line—CO component. $V_\infty = 5223$ m/s, $\rho_\infty = 2.933 \times 10^{-4}$ kg/m^3.

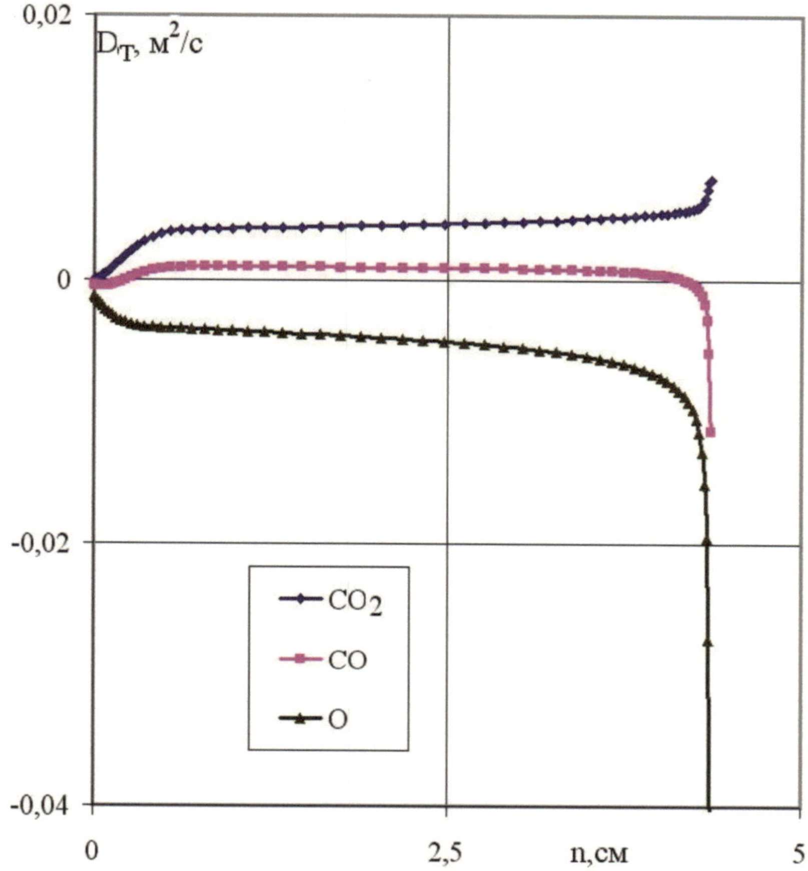

Figure 8. Distribution of thermo-diffusion coefficient D_T along the stagnation line. $V_\infty = 5223$ m/s, $\rho_\infty = 2.933 \times 10^{-4}$ kg/m³.

The Schmidt number is characterized the ratio of processes of momentum and mass transfer. For multi-component gas mixtures Schmidt's, Lewis's numbers depend on temperature and species fraction. For multi-component gas mixtures, Schmidt's numbers are defined for every pair of gas mixture. In practice during numerical calculations of chemically non-equilibrium flow, Schmidt's number are chosen be equal to constant. Sometimes to all components of a mixture Schmidt's number is used as identical. In this connection, it is important to estimate the influence of this supposition on the received results. Let us remind the definition of Schmidt's number $S_{ij} = \mu/\rho D_{ij}$, $D_{ij} = C_3C_4/pQ_{ij}^{1,1}$, $C_3 = 8.256{*}10^{-7}T^{3/2}$, $[D_{ij}] = m^2/c$, $[p] = atm$, $Q_{ij}^{1,1} = A^{02}$.

The Schmidt's number distribution along the stagnation line are shown in **Figure 11**.

Distribution of Lewis numbers along the stagnation line (Le = Pr/Sc, Pr is the Prandtl's number) for one of variants resulted in **Figure 12**. We shall notice, that near the surface of the vehicle ($n = 0$), Lewis's number considerably differs from unit that testifies discrepancy of mass velocities due to heat conductivity and diffusion in this area.

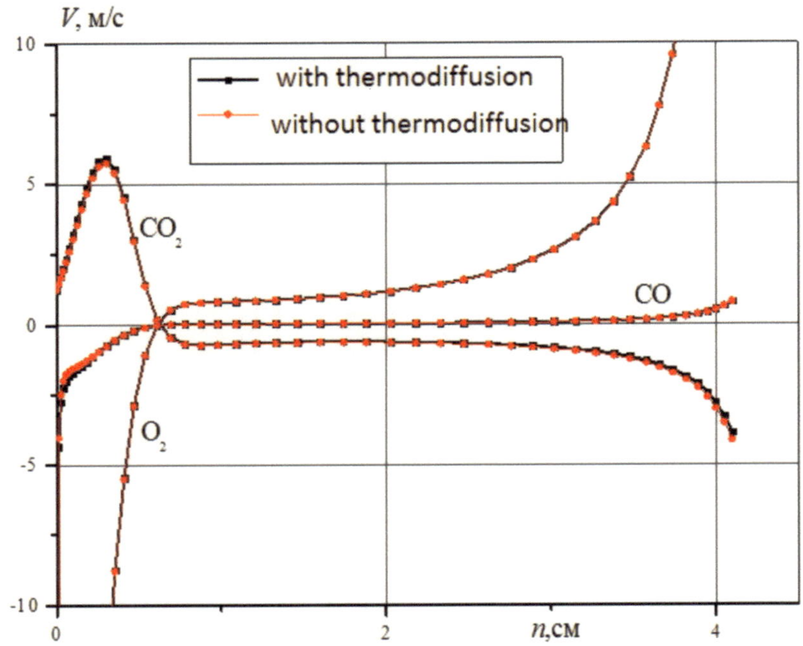

Figure 9. Distribution of diffusion velocity of different mixture component along the stagnation line with influence and without thermo-diffusion. $V_\infty = 5223$ m/s, $\rho_\infty = 2.933 \times 10^{-4}$ kg/m^3.

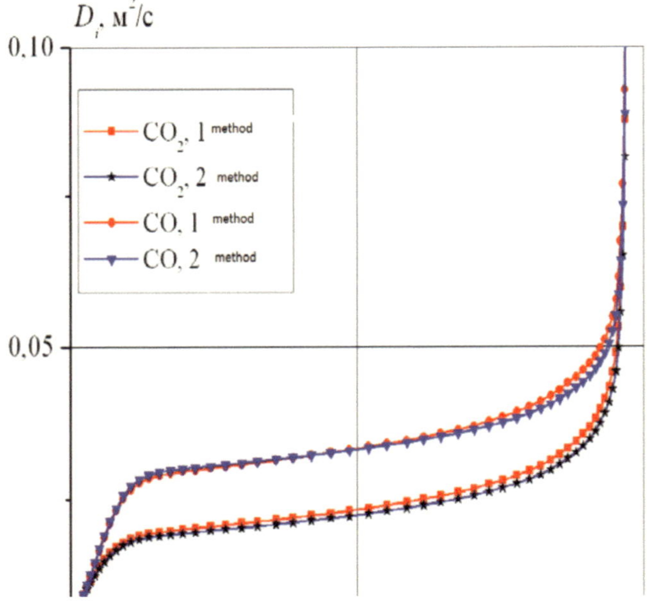

Figure 10. Effective diffusion coefficient along the stagnation line for CO_2 and CO, is obtained by two methods: $V_\infty = 5223$ m/s, $\rho_\infty = 2.93 \times 10^{-4}$ kg/m^3.

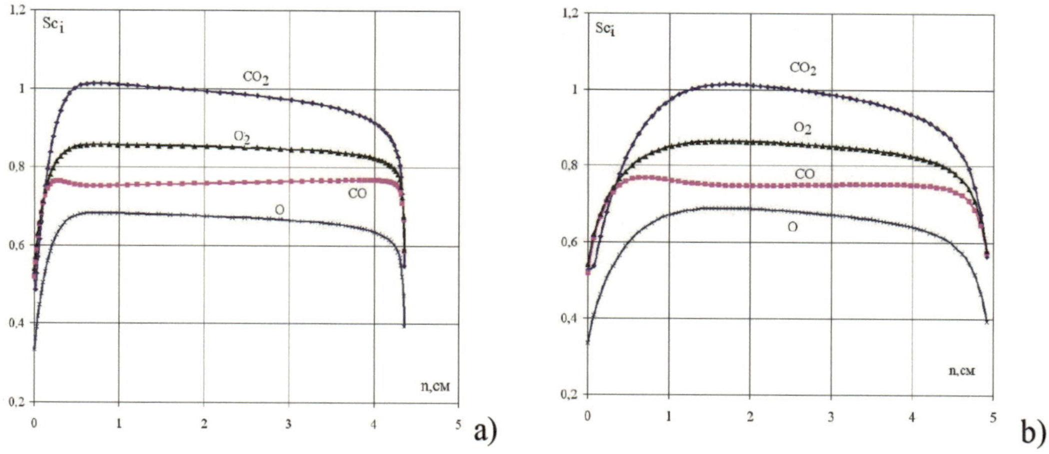

Figure 11. Values of Schmidt number along the stagnation line, ideal catalytic surface, (a) $V_\infty = 5223$ m/s, $\rho_\infty = 2.933 \times 10^{-4}$ kg/m^3, (b) $V_\infty = 5687$ m/s, $\rho_\infty = 3.141 \times 10^{-5}$ kg/m^3.

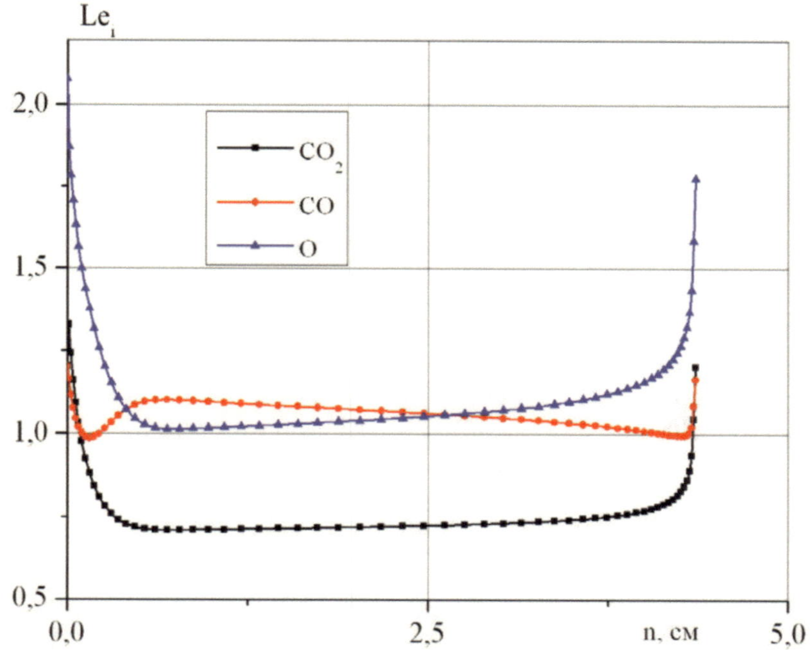

Figure 12. Values of Lewis's number along the stagnation line for different component of mixture, ideal catalytic surface. $V_\infty = 5223$ m/s, $\rho_\infty = 2.933 \times 10^{-4}$ kg/m^3.

5. Boundary conditions

The solution must be found out in the region restricted by: (a) body surface; (b) inflow: surface of external flow, where the conditions are known—$V_\infty, p_\infty, \rho_\infty, c_{i\infty}$; (c) axis of symmetry:

symmetrical or anti-symmetrical reflection depending on functions; (d) outflow: some surface in down part of flow, where usually ones use so called "soft" boundary extrapolation conditions.

The boundary conditions at the thermo-chemically stable surface include no slip conditions for component of velocities. Scott's wall slip conditions applied to velocity, species mass fractions for modeling flow fields in high altitude [25].

Appropriate boundary conditions at thermally stable surface include conditions for the diffusive fluxes of element at the wall, mass balance equations for the reaction product. When the temperature of the wall is done ($T = T_\infty$), then boundary conditions at the surface include L conditions for the elemental diffusive fluxes at the wall

$$J^*_{jw} = 0, (j = 1, \dots, L) \tag{26}$$

and the mass balance equations for the reaction products

$$J_{iw} = (\rho_w c_{iw})^\nu K_{wi}, (i = L + 1, \dots, N), \tag{27}$$

where $K_{wi} = \gamma (kT_w / 2\pi m_i)^{1/2}$ is the effective catalytic constant. Here recombination is qualitatively characterized by an effective probability $0 < \gamma < 1$ or by rate constant K_{wi}($K_{wi} = 0$ for an non-catalytic wall, $K_{wi} = \infty$ for a perfect catalytic wall). Value ν is the order of the reaction, mi is the atomic mass. The cases of $\gamma = 1$ and 0 correspond to absolutely catalytic and absolutely non-catalytic materials. The catalytic property of the wall has an important effect on the heat transfer of reusable vehicles over the considerable interval of the trajectory. The structure of the surface (contamination, roughness, porosity, etc.) affects the rates of the atomic adsorption and de-sorption processes.

A phenomenological model for catalytic reactions used that accounts for physical and chemical absorption, the interaction between the impinging atoms and ad-atoms (adsorbed atoms), and between the ad-atoms themselves. A model of the Rideal-Eley and Langmuir-Hinshelwood layer with ideal adsorption applied. Let us consider the heterogeneous catalytic reactions on surfaces [6]:

$$O + O \rightarrow O_2, ⊚O + CO \rightarrow CO_2 \tag{28}$$

For a surface with final catalytic properties, it is applicable the simplified boundary conditions with use of effective probabilities of heterogeneous recombination that are equal among themselves $\gamma_O = \gamma_{CO} = \gamma_w$. Diffusion fluxes on a surface for a molecule CO and atoms O can be written as follows:

$$-J_{CO} = \rho k_{wCO} c_{CO}, \; -J_O = \rho k_{wO} c_O, k_{wi} = \frac{2\gamma_w}{2 - \gamma_w} \sqrt{\frac{RT_w}{2\pi m_i}}, i = CO, O \tag{29}$$

And for molecules O_2 and CO_2

http://dx.doi.org/10.5772/intechopen.71666

$$J_{O_2} = -J_O, J_{CO_2} = -J_{CO} - J_O \tag{30}$$

Above boundary condition for the mass concentration on the body surface can be expended

$$\rho D_{ij} \partial c_i / \partial y = (\rho_w c_{iw})^\nu K_{wi} \tag{31}$$

If surface posses by catalytic properties then surface provoke to recombine the atoms in molecules. On absolutely catalytic surface concentration of atoms, it is equal to zero: $c_A = 0$. For absolutely non-catalytic wall—$\partial c_A / \partial y = 0$.

An expression for the heat flux to the surface may be deduced (for simplicity, relaxation is considered to be already completed at the wall), that is.

$$J_{qw} = -\lambda \partial T / \partial y - \sum_{k=L+1}^{N} h_k J_k \tag{32}$$

The heat flux depends essentially on the boundary conditions for the species concentrations at the wall. Thus at $K_{wi} = 0$, we obtain

$$J_{qw} = -\lambda \partial T / \partial y \tag{33}$$

For reusable vehicles, the catalyze quality of heat-protective coverings become very important. The heat flux increases as the diffusion contributes to the maximum total heat flux. Since homogenous recombination and neutralization occur slowly at high altitudes, exothermic heterogeneous processes at the body surface become crucial to the magnitude of the convection heat flux.

Next equation may be used to find the temperature with the boundary condition of heat balance at the wall between flux to surface and reflection. The energy equation yields

$$J_{qw} = \varepsilon \sigma T_w^4, \tag{34}$$

where ε is a measure of the surface blackness and σ *is* the Stefan-Boltzmann constant.

Rotational temperatures of molecules are equal to the translational temperature of heavy atomic particles due to a fast translational-rotational energy exchange requiring only several collisions to establish the Boltzmann distribution. In the free stream, CO_2 molecules have almost zero vibration energy, therefore, for them in a shock layer, there is an area with non-equilibrium vibration. Vibration temperatures of all the electronically exited molecules are considered to be equal to the translation temperature of heavy atomic particles.

6. Mathematical models and numerical methods

Hypersonic flows over real space configurations represent a substantial problem from the point of view of the development of new and more effective mathematical models, numerical algorithms and the use of modern computer systems.

During the past decade, a large number of computational codes have been developed that differ in the grid generation methods and numerical algorithms used. For numerical simulation of external flow fields, past real form bodies are necessary to construct the geometry, to design a discrete set-grid, to provide the mathematical model of the initial value problem, to approximate the governing equation by numerical ones, to design a computational algorithm, to realize the flow field, to establish a feed-back of obtained results with experiment, analytical and benchmark problems, and so on.

As mathematical model, the Navier-Stokes equations and the various sub-models obtained in frameworks of the asymptotic analysis sub- and supersonic flow past blunted bodies in various statements and in a wide range of numbers of Reynolds are used.

Traditional asymptotic analysis of Navier-Stokes equations for different regimes of viscous compressible flow depending on small parameter 1/Re make it possible to decouple the different types of gas flows. The next methods were used: Navier-Stokes equations in so-called approximation of a viscous shock layer and full Navier-Stokes (N-S) equations. For solution of governing equations, the implicit finite-difference monotone schemes of the second order are used [15, 16]. Generalized Rankine-Hugoniot's conditions are imposed in the shock wave. Special method of high stiffness resolution of non-equilibrium phenomena is applied [16].

The Navier-Stokes equations are written in a conservative form for arbitrary coordinate system. The implicit iterative scheme is based on a variant of Lower-Upper Symmetric Gauss-Seidel (LU-SGS) scheme. At high altitudes (low Reynolds numbers) where the bow shock has a finite thickness, a shock capturing approach is used. So inflow boundary conditions are specified in the free stream. At lower altitudes, a shock fitting scheme is employed with the modified Rankine-Hugoniot conditions specified at the bow shock. Besides the Navier-Stokes equations at lower altitudes, the viscous shock layer equations are also solved. This implicit scheme leads to the scalar diagonal manipulation for a case of non-reacting perfect gas flow and does not require any time-consuming matrix inversion. In more details, the numerical methods is described in [26–28] for the shock layer equations.

7. Results of numerical investigation of hypersonic flow past space vehicle in Martian atmosphere

The investigations of a hypersonic flow past a frontal part of MSRO (Mars Sample Return Orbiter) and MARS EXPRESS vehicles descending in an atmosphere of Mars are shown below. The hypersonic Mach number means that appreciable quantity of molecules in high-temperature region began to dissociate. For an Earth atmosphere such numbers is equal to $M_\infty \geq 6$. For an atmosphere of Mars in which main component is carbon dioxide as hypersonic numbers, the values $M_\infty \geq 10$ are considered. The typical regimes of the entry in a Martian atmosphere are considered (**Figure 2**). The conditions of a flow corresponding to the last stage of flight of space vehicles in an atmosphere of Mars ($V_\infty \leq 6$ km/s, $\rho_\infty \approx 10^{-5}$ kg/m^3, $H < 60$ km) were studied. Determining process at such velocities is a process of dissociation. Up to 75% of full gas flow, energy can be spent on it [29].

The region where non-equilibrium physical and chemical processes realized is a significant part from all considered regions. Velocity of physical and chemical processes, as a rule, grows together with density of gas. As the density of an atmosphere of Mars is much less than in atmosphere of the Earth, the equilibrium flows for bodies of the moderate sizes are observed at smaller altitude: H <10–20 km—for an atmosphere of Mars, H ≤ 30 km—for an atmosphere of the Earth.

At high temperatures that observed in a shock layer, the characteristic times of a vibration energy relaxation of molecules and characteristic times of dissociation become one order. Thus the account of non-equilibrium excitation of vibration degrees of freedom of carbon dioxide molecules is necessary.

7.1. Some features of a reacting gas mixtures flow

At a supersonic flow, the main features of reacting gas mixture can be evidently shown by change of flow parameters across shock layer. The distribution of pressure, velocities in a shock layer depends on physical and chemical processes weakly. The pressure with high degree of accuracy is estimated in limits between values $p = \rho_\infty V_\infty^2(1-k)$ behind a direct shock wave and $p = \rho_\infty V_\infty^2(1-0.5\cdot k)$ in a stagnation point [29]. Here value $k = \rho_\infty/\rho_s$ is the characteristic value of gas compression in the shock layer equal the ratio of density in an external flow and density behind a direct shock wave. For flow parameters of MARS EXPRESS vehicles presented in **Table 1**, the pressure in a stagnation point equals to values 0.95–0.96 of a high-speed pressure $\rho_\infty V_\infty{}^2$. We shall notice that for the perfect gas with a parameter of an adiabatic ratio γ = 1.4 at the given velocities, the pressure in a stagnation point takes ~0.92 from a high-speed pressure.

Main results are shown: (1) in shock layer across of stagnation line; (2) along of surface body for heat transfer; and (3) in shock layer along body. We used the orthogonal system of coordinates (ξ, ζ). One coordinate ξ directs from a forward stagnation point along a streamline contour along the surface. The coordinate ζ is a normal to wall.

The change of specific heat capacity ratio $\gamma = c_p/c_v$ (c_p is the specific heat capacity at constant pressure and c_v is the specific heat capacity at constant volume) is shown in **Figure 13**.

H, km	V_∞, m/s	ρ_∞, kg/m^3	T_∞, K	Re_∞	$P_0/(\rho_\infty V_\infty{}^2)$
52.59	5923	7.61×10^{-5}	140	1.7×10^4	0.96
43.01	5292	2.51×10^{-4}	149	5.0×10^4	0.96
36.16	4259	5.58×10^{-4}	158	9.1×10^4	0.95
32.42	3433	8.45×10^{-4}	163	1.1×10^5	0.96

Table 1. Trajectory parameters of MARS-EXPRESS.

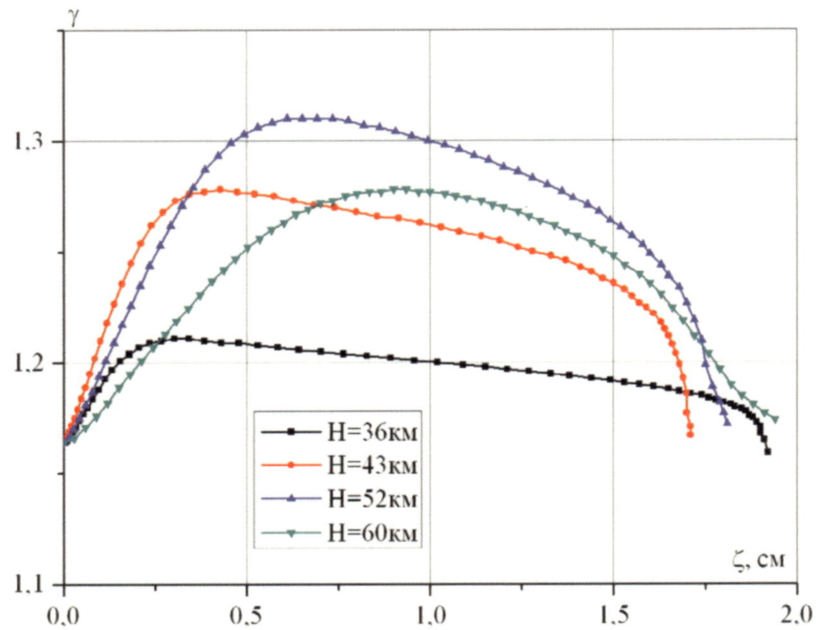

Figure 13. Changing ratio of specific heat capacity for flows past MSRO.

The ratio is always greater than 1 and its value is an important indication of the atomicity of the gas. The laminar-to-turbulent transition of flow on the frontal surface proceeds at the altitude below 20 km. Thus the taking into account of the boundary layer transition does not affect on the results of heating the thermal protection. In **Figure 14**, distributions of shock layer temperature along a stagnation line near spherically blunted body (radius R = 1 m) under various conditions of a flow are shown. It includes the regimes from completely viscous shock layer until flow with a thin boundary layer. Parameters of flow in a shock layer is obtained in approach of a viscous shock layer by numerical computation and with help of the physical and chemical models submitted in work [10]. Thus Reynolds's number—Re_∞ varied (due to change of density of an external flow) from 5×10^3 to 1.5×10^5. In the shock layer, the pressure determined by the velocity and density of the external flow equal to $p = 0.22$ atm ($Re_\infty = 1.5 \times 10^5$) and in the most part of a shock layer close to equilibrium value (a curve 4).

With reduction of pressure and also Reynolds's numbers the length of non-equilibrium region increases and at p = 0.007 atm character of flow in a shock layer becomes closer to frozen (curve 1). For comparison in **Figure 14**, the structure of temperature on a stagnation line of a flow without taking into account physical and chemical transformations is shown also (curve 5). This calculation is carried out under condition of laws of the perfect gas. It is visible that behind a boundary layer, the temperature leaves on "slop" and its value in some times higher than in case of a flow with chemical reactions. The estimations show that in case of chemical reactions under the given conditions up to 75% of full energy of an external flow it can be spent on dissociation molecules. As a result, the temperature in a shock layer essentially goes down and the density of gas increases. Thus the share of kinetic energy spent on realization of

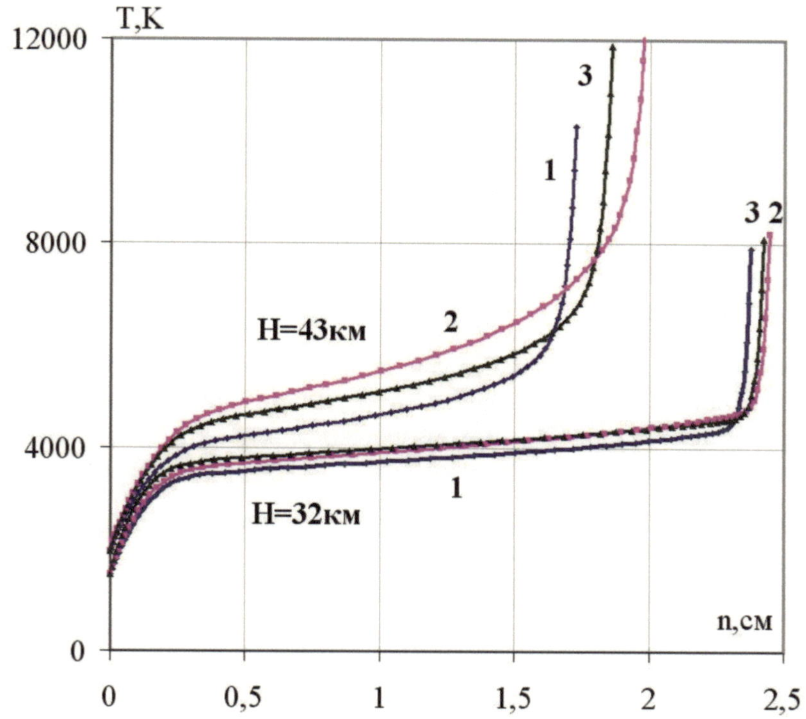

Figure 14. Distribution of temperature along stagnation line for two points MARS EXPRESS vehicle. 1: model of [30], 2: model of [31], 3: model of [32–34].

physical and chemical processes in a shock layer is essential that affects distribution of temperature and does inapplicable many results of gas dynamics of the perfect gas.

7.2. Influence of various models of chemical kinetics on a hypersonic flow past bodies

One of the most important problems of a hypersonic flow is the account of the real physical and chemical transformations in a shock layer. In theoretical works, the different authors used models of the chemical reactions essentially differing by reaction rate constants. Let us carry out comparison of the basic models used for calculation of chemical reaction rate constants in a high-temperature flow of carbon dioxide gas. We can estimate the influence of these models on character of a flow and heat transfer to the wall. The basic models of chemical reactions have essentially different reaction rate constants in a high-temperature flow of carbon dioxide gas. The corresponding dissociation reaction rate constants in a considered range of temperatures can differ up to two orders in dependence on used models [30–35].

In works [10, 13, 17–19], numerical research of a non-equilibrium flow of the bodies modeling the form of Martian vehicles MARS EXPRESS and MSRO with use of these models is carried out. The surface of the vehicles was considered or as ideal catalytic (the maximal velocities heterogeneous recombination a component of dissociated carbon dioxide gas), or non-catalytic (velocities of heterogeneous recombination of component is equal to zero). We shall consider some results of numerical researches.

As an example of the distribution of CO_2 concentration in **Figure 15** in shock layer for two type of cone; (a) $\theta = 60°$, (b) $\theta = 10°$; $V_\infty = 5223$ m/s, $\rho_\infty = 2.93 \times 10^{-4}$ kg/m^3 is shown.

In **Figure 16**, structures of mass concentration component CO_2 and CO obtained in case of use one-temperature ($T_v = T$) and two-temperature [36] reaction rate constants of chemical reactions. It is evident that for considered flow conditions an influence of non-equilibrium vibration on dissociation process is insignificant: structures of mass concentration CO_2 and CO coincide almost in all region of a shock layer. The small divergence is observed near a shock wave. From data, it is followed that process vibration non-equilibrium does not affect on parameters of flow near a body.

With use of these models, chemically non-equilibrium flow is considered and their influence on parameters of flow and heat exchange is established. The significant differences in distributions of temperature, concentration of a component of a gas mixture in a shock layer is observed at a variation of model chemical kinetics.

From the data shown in **Figure 17**, it is evident that for conditions of flight $H = 43$ km, the significant differences of values of mass concentration of carbon dioxide and an withdrawal of the shock wave from a surface is observed. For $H = 32$ km, corresponding values practically coincide. This fact can be explained at an altitude $H = 43$ km, the mode of flow in a shock layer is far from chemical equilibrium. In this case, the parameters of flow depend on reaction rate constants of direct and reverse chemical reactions. For different models under these conditions of a flow, it differs essentially. Therefore in considered case, if you used models of Park then the dissociation reaction CO_2 in the disturbed region goes with much more rate than it is proposed by another two models. The chemical components in a shock layer affects on distributions of temperature and also on size of a withdrawal of a shock wave from a surface of a body.

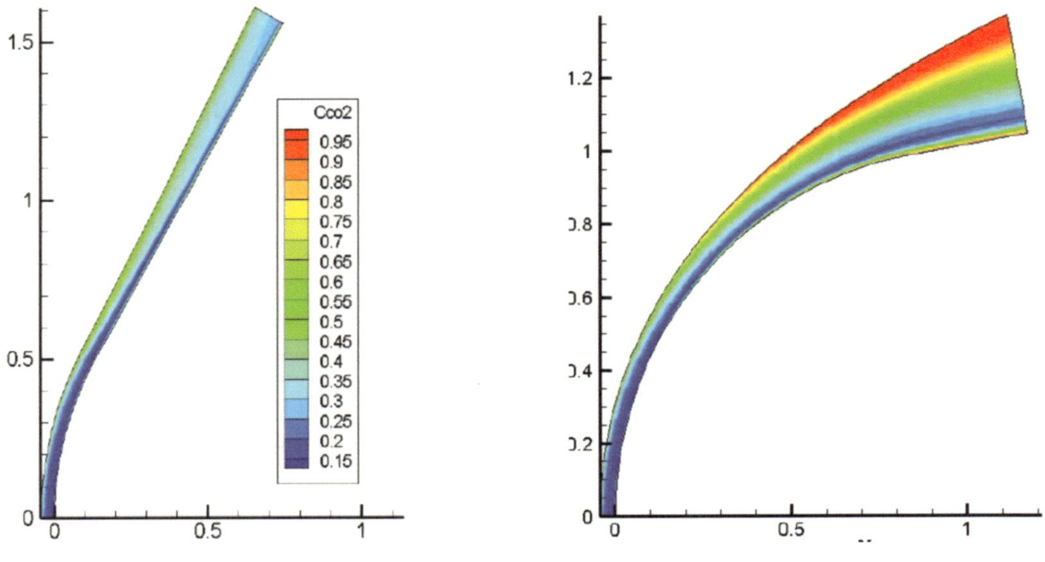

Figure 15. Distribution of CO_2 concentration in shock layer for two type of cone; (a) $\theta = 60°$, (b) $\theta = 10°$; $V_\infty = 5223$ m/s, $\rho_\infty = 2.93 \times 10^{-4}$ kg/m^3.

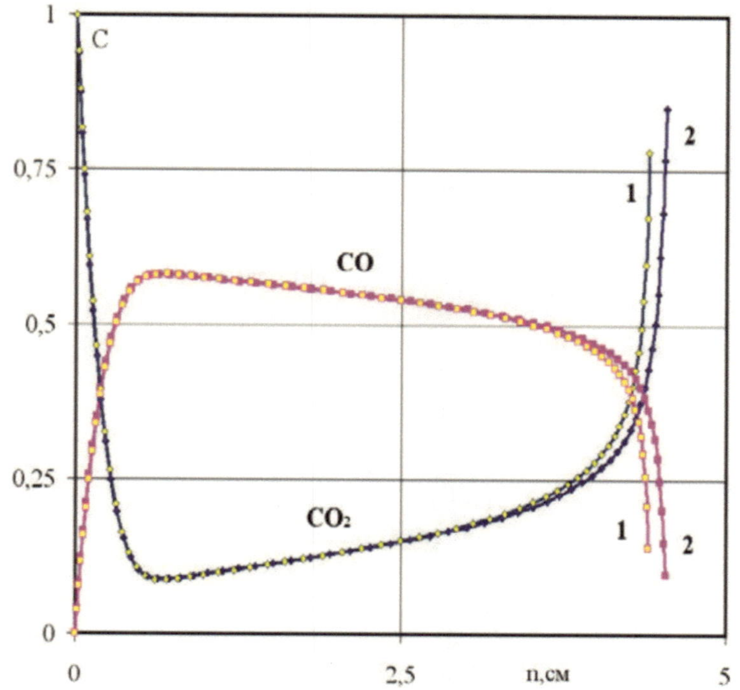

Figure 16. Distribution of mass concentration CO_2 and CO along the stagnation line, 1—with, 2—without taking into account the influence of vibration relaxation on process dissociation, $V_\infty = 5223$ m/s, $\rho_\infty = 2.93 \times 10^{-4}$ kg/m^3.

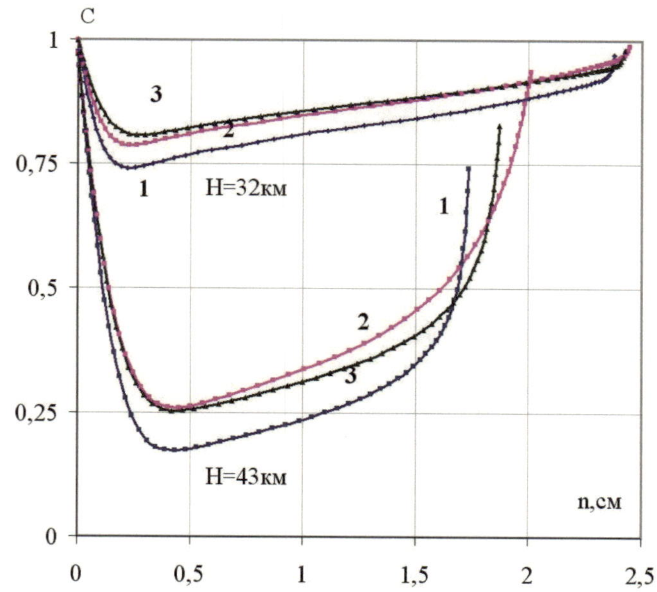

Figure 17. Profile of mass concentration of CO_2 along the stagnation line for two point of trajectories MARS EXPRESS vehicle. 1: model of [30], 2: model of [31], 3: model of [32–34].

With reduction of velocity flight, it corresponds with an increase of density in a shock layer and flows more close to equilibrium case. The parameters of flow and structure of a gas mixture are defined by conditions of chemical equilibrium. The reaction rate constants in equilibrium will be the same for all used models. In this connection, corresponding structures of temperature and concentration, and also value of a withdrawal of a shock wave from a surface received for different models under conditions of a flow at height H = 32 km well enough coincide.

Influence of chemical models on flow parameters and heat exchange is determined in a wide range of parameters of a flow of MSRO vehicle in cases ideally catalytic and non-catalytic wall. In **Figures 18** and **19**, mass concentrations of component CO_2 and CO along stagnation line for different models and types of condition on wall are submitted. The data are resulted for conditions of flow $V_\infty = 5687$ m/s, $\rho_\infty = 3.14 \times 10^{-5}$ kg/m^3 that corresponds approximately to altitude of flight $H \approx 60$ km; the temperature of a surface equal to $T_w = 1500$ K.

Comparing the data of **Figures 18** and **19**, it is possible to notice that influence of catalytic surface properties affects profiles of concentration basically near to a wall. In these figures, the significant divergence in the distributions of concentration the components obtained for different models is observed. It is evident that it is greater for degree CO_2 dissociation in a shock layer when it used of model of Park [31], and is smaller—when the model of Kenzie-Arnold [30] used. The model of Research Institute of Mechanics (NIIMekh) of Moscow State University [32–34] gives the intermediate results. It is established that change of reaction rates practically does not influence on value of a heat flux to ideal catalytic wall of the vehicle. In a

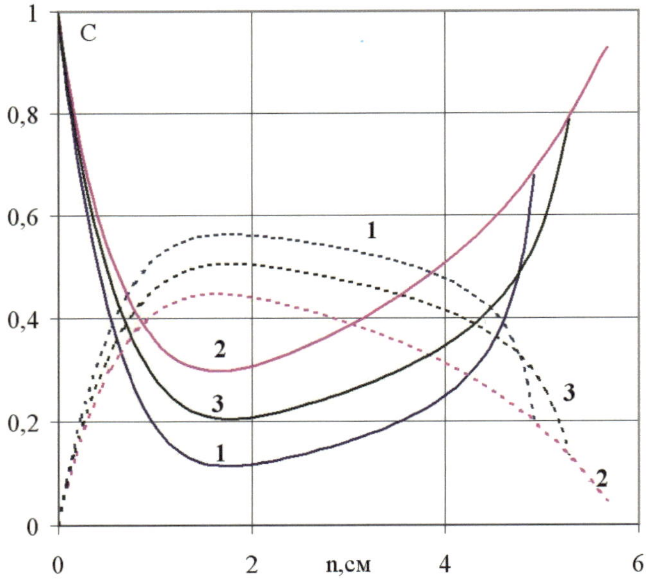

Figure 18. Profiles of mass concentration CO_2 (——) and CO (- - - -) along stagnation line MSRO vehicle, ideal-catalytic wall, 1: model of [30], 2: model of [31], 3: model of [32–34].

case non-catalytic surfaces the difference in values of the heat fluxes received for different models can be essential up to 30%.

In **Figure 20**, for two points of a trajectory of MARS EXPRESS vehicle (altitude of flight equal to $H = 43$ km and $H = 32$ km) the distributions of temperature obtained with the help of model

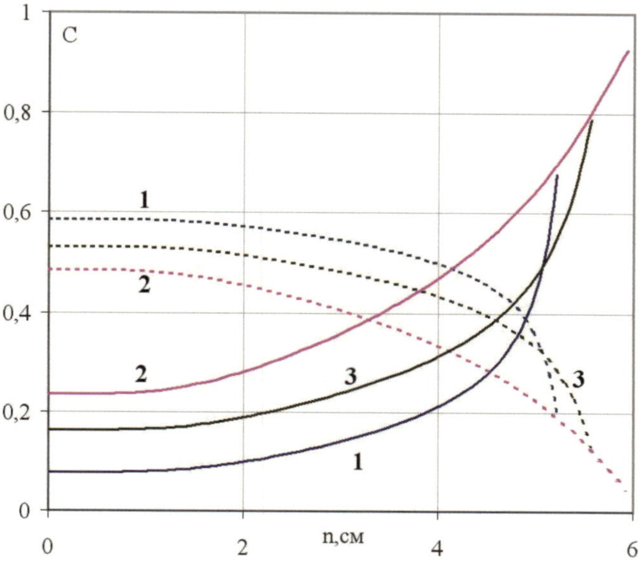

Figure 19. Profiles of mass concentration CO_2 (——) and CO (- - - -) along stagnation line MSRO vehicle, non-catalytic wall, 1: model of [30], 2: model of [31], 3: model of [32–34].

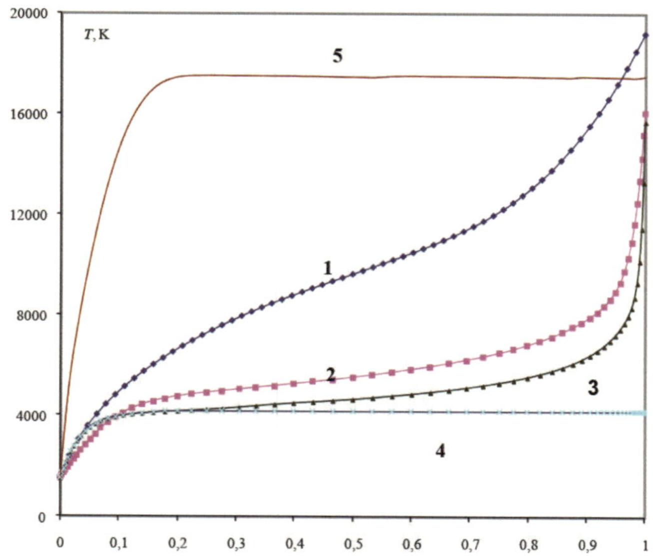

Figure 20. The temperature along the stagnation line for sphere $R = 1$ m in Mars atmosphere. 1: pressure $p = 0.007$ atm, $Re_\infty = 5 \times 10^3$; 2: 'frozen' flow: $p = 0.07$ atm, $Re_\infty = 5 \times 10^4$; 3: $p = 0.22$ atm, $Re_\infty = 1.5 \times 10^5$; 4: $p = 0.22$ atm, $Re_\infty = 1.5 \times 10^5$, equilibrium case; 5: $p = 0.065$ atm, $Re_\infty = 1.3 \times 10^5$, perfect CO_2 gas.

of Park [31], models of Kenzie-Arnold [30] and the model developed in Research Institute of Mechanics (NIIMekh) of Moscow State University [32–34] are shown. The data are resulted along a stagnation line. The surface of the vehicle it is an ideal—catalytic wall with the constant temperature.

Figure 21 shows the density distribution along the stagnation line for three conditions of a flow of MSRO vehicle in an atmosphere of Mars. In case of non-equilibrium flows, the density along stagnation line considerably changes.

7.3. Influence of non-equilibrium vibration kinetics on parameters of a flow

Let us consider results of numerical calculation of a non-equilibrium flow from point of view of different vibration relaxation models. The influence of non-equilibrium excitation of vibration degrees of freedom of carbon dioxide was investigated on a basis of three-temperature kinetic model and two simplified case: in two-temperature approach when it introduced uniform vibration temperature for all types of vibration of CO_2 molecule and in one temperature approach when translational and vibration temperature was the same [10, 36].

Structures of two specific vibration energy E_{12} and E_3 for conditions of a flow past MSRO vehicle $V_\infty = 5223$ m/s, $\rho_\infty = 2.93 \times 10^{-4}$ kg/m^3 (altitude of flight ~40 km) along the stagnation line in shock layer obtained with use of three-temperature model of a vibration relaxation and presented in **Figure 22**.

These profiles are characterized by significant flow gradients in relaxation region near a shock wave and their behavior reflects of features of considered vibration relaxation model of molecules CO_2 which is taking into account internal structure of molecules CO_2 and extra mode exchanges by vibration energy.

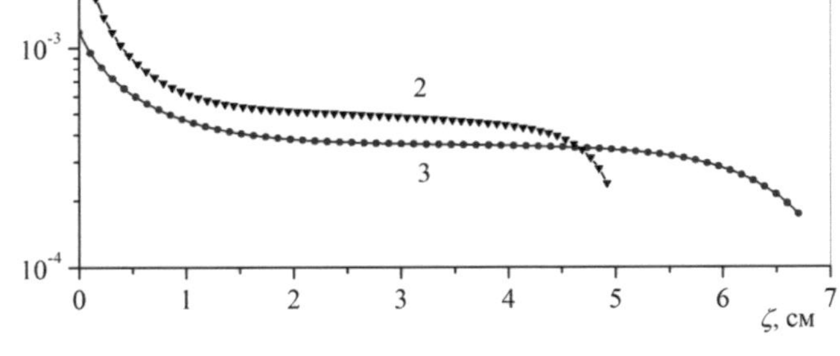

Figure 21. Density distribution along the stagnation line for different condition of flow past MSRO vehicle. 1: $V_\infty = 5223$ m/s, $\rho_\infty = 2.933 \times 10^{-4}$ kg/m^3; 2: $V_\infty = 5687$ m/s, $\rho_\infty = 3.141 \times 10^{-4}$ kg/m^3; 3: $V_\infty = 3536$ m/s, $\rho_\infty = 2.819 \times 10^{-5}$ kg/m^3.

In **Figure 23**, comparison of total rotation energy $E_v = E_{12} + E_3$ and the vibration energy E_v obtained with the help of three-temperature [10] and two-temperature models [36] for two conditions of a flow of the vehicle is carried out: $V_\infty = 5223$ m/s (altitude of flight H ~ 40 km) and $V_\infty = 5687$ km/s (H ~ 60 km).

The vibration energy difference near a shock wave in relaxation zone is visible. The structures received with the help of two-temperature model have more "smearing" type than the structures

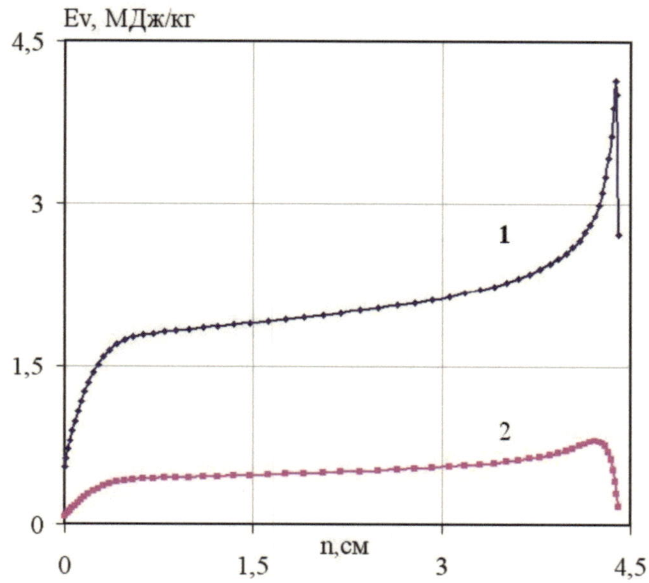

Figure 22. Distribution of vibration energy component along stagnation line MSRO vehicle, three-temperature model [10], 1: E_{12}, 2: E_3. $V_\infty = 5223$ m/s, $\rho_\infty = 2.93 \times 10^{-4}$ kg/m^3.

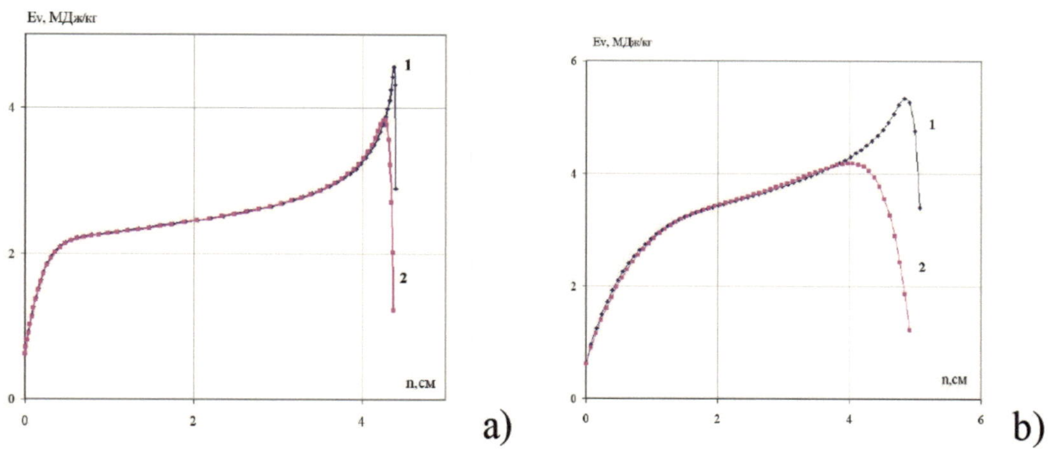

Figure 23. Distribution of vibration energy along stagnation line MSRO vehicle; 1: $E_v = E_{12} + E_3$, three-temperature model of [10]; 2: E_v, model of [37]; (a) $V_\infty = 5223$ m/s, $\rho_\infty = 2.93 \times 10^{-4}$ kg/m^3; (b) $V_\infty = 5687$ m/s, $\rho_\infty = 3.141 \times 10^{-5}$ kg/m^3.

received with use of more rigorous three-temperature approach. You can see also that the maximum values of vibration energy obtained with the help of three-temperature model on 15–20% surpass corresponding values of two-temperature model. In the most part of a shock layer, conditions of thermodynamic equilibrium are realized and both models give identical result.

It is possible to estimate the non-equilibrium vibration zone value considering the structures translational and vibration temperatures T_{12}, T_3. The various modes received with help of three-temperature model and for two variants of a flow are displayed in **Figure 24**. It is visible that the size of this zone is more for the second variant of a flow (V_∞ = 5687 m/s) corresponding to altitude ~ 60 km.

Comparison of values of translational temperature for two conditions of a flow in the assumption of a weak deviation from thermal equilibrium at $Tv = T$ (a curve 1) and in non-equilibrium gas (curves 2, 3) along a stagnation line is presented in **Figure 25(a)** and **(b)**. Corresponding to relaxation models profiles of temperature (curves 2 and 3) for these conditions practically coincide, a divergence no more than 5%.

It is visible that the account of non-equilibrium excitation of vibration degrees of freedom of molecules CO_2 leads to insignificant increase in a withdrawal of a shock wave from a surface of a body and to essential increase (~ on 25–30%) translational temperature in the field of a shock wave in comparison with thermally equilibrium case. It does not influence on a gas mixture temperature near to a body surface. The fact of translational temperature increase in relaxation zone is connected with transition of internal degrees energy of freedom of molecules CO_2 in translational energy of others components.

7.4. Processes of heat transfer in the multi-component mixture

Comparison of a heat transfer in the assumption thermally equilibrium (curves 1) and non-equilibrium gas (curves 2, 3) in cases ideal catalytic (**Figures 26** and **27**) and non-catalytic

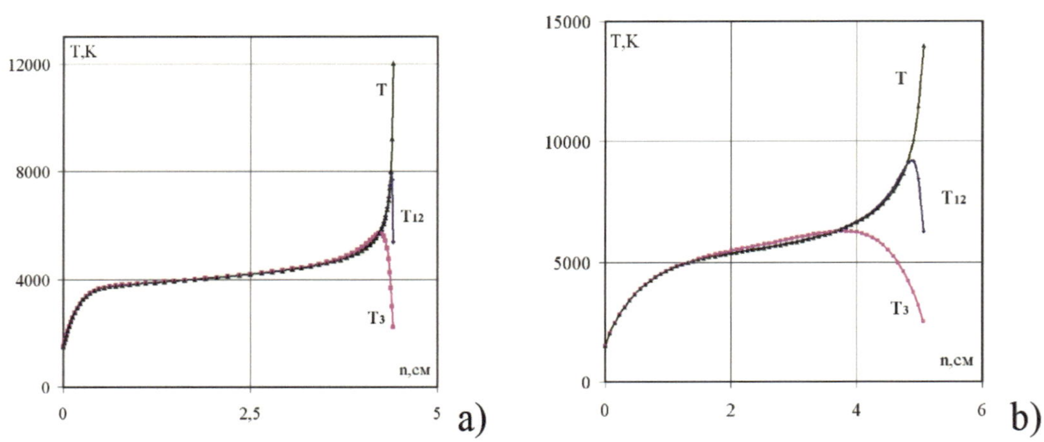

Figure 24. Profiles translational (T) and vibration (T_{12}, T_3) temperature for MSRO vehicle, three-temperature model of [10]; (a) V_∞ = 5223 m/s, ρ_∞ = 2.93 × 10^{-4} kg/m^3; (b) V_∞ = 5687 m/s, ρ_∞ = 3.141 × 10^{-5} kg/m^3.

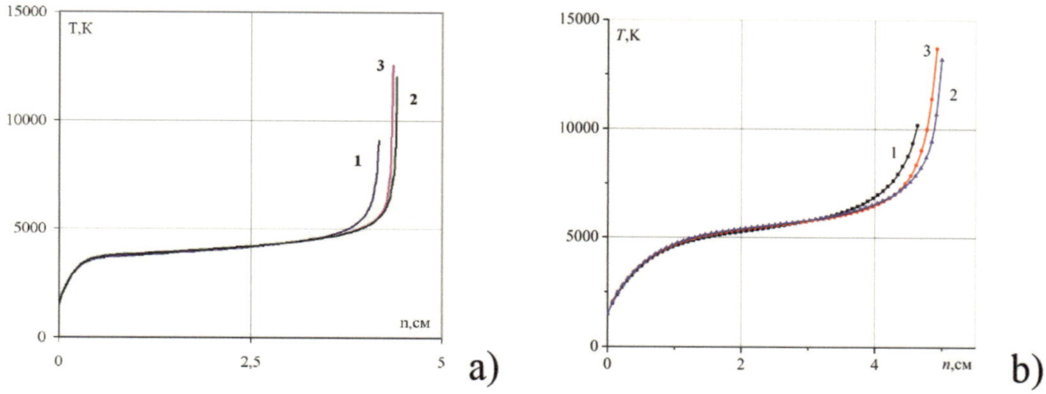

Figure 25. The translational temperature along the stagnation line for MSRO vehicle, 1: equilibrium flow, 2: non-equilibrium flow, three-temperature model of [10], 3: non-equilibrium flow, two-temperature model [38], (a) V_∞ = 5223 m/s, $\rho_\infty = 2.93 \times 10^{-4}$ kg/m^3, (b) V_∞ = 5687 m/s, $\rho_\infty = 3.14 \times 10^{-5}$ kg/m^3.

surfaces (**Figure 27**) is carried out. It is visible that the thermal flux from point of view of non-equilibrium excitation of vibration degrees of freedom of molecules CO_2 can surpass on 10% a flux for thermally equilibrium gas.

Comparing the data in **Figures 26** and **27** for the same conditions of a flow, we shall notice that the heat transfer to ideal-catalytic surface in 3–4 times surpasses a corresponding flux to non-catalytic wall.

This fact can be explained by that in a case when on a surface the maximal rates of heterogeneous recombination of a component of carbon dioxide gas is observed (ideal catalytic), the chemical energy spent on dissociation is allocated and transferred to a surface. If recombination does not occurs (low catalytic activity), gas components pass down flow, carrying away with itself the dissociation energy and the additional heat transfer does not occur.

As it was already observed, the heat transfer to ideal catalytic wall does not depend on used model of chemical reaction rates that the curves show in **Figure 26**. For non-catalytic walls, the heat transfer to a surface with using of Kenzie-Arnold model [30] approximately on 30% exceeds a heat transfer obtained with help of model of Park [31]. It shows that at non-catalytic surfaces of the vehicle does not occur of recombination of a component. In this case, reactions go only in one direction, and with different velocities for different models. As result for considered models, the different chemical composition of a gas mixture near surfaces (**Figure 29**) obtained. Also divergences in values of a heat transfer are observed. Besides the recombination reactions on a surface of additional body heating does not occur. Therefore value of a heat transfer for non-catalytic walls in 3–3.5 times less than corresponding values for ideal catalytic surfaces (**Figure 27**). A heat transfer distribution of along a frontal surface of MARS EXPRESS vehicle (it is ideal catalytic surface) with use of three models chemical kinetics is resulted in **Figure 29**. In spite of various rates of reactions, the good correlation of results observed for a heat transfer to a surface for considered models. Really in a case, ideally catalytic wall last plays a role of the catalyst. It is promotes the reactions of recombination and as a result the

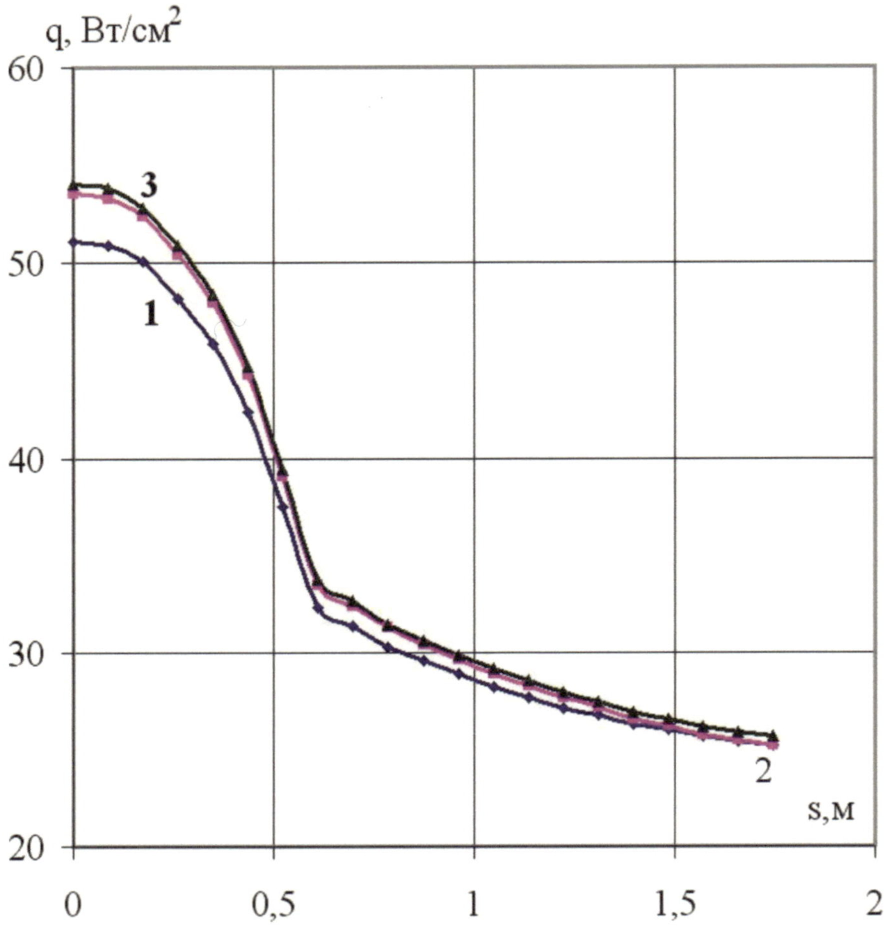

Figure 26. Heat transfer to MSRO vehicle surface in case of ideal catalytic wall; 1: equilibrium flow, 2: non-equilibrium flow for three-temperature model [10], 3: non-equilibrium flow for two-temperature model [38]. V_∞ = 5223 m/s, $\rho_\infty = 2.93 \times 10^{-4}$ kg/m^3.

chemical energy spent on dissociation is transferred to wall. This process is determined mainly by conditions of equilibrium of reactions.

Influence of non-equilibrium excitation of vibration degrees of freedom of molecules CO_2 on distribution of a heat flux along a surface of a body is displayed in **Figures 26–29**.

In **Figure 29** for comparison (a triangular marker) the results of works [6] in which Navier-Stokes equations solved by method of finite volume are submitted also. In this case, the model of chemical reactions [32–34] was used. For some conditions of a flow the discrepancy reached (up to 20%) for the obtained values of heat fluxes. The corresponding data of [6] obtained for different boundary conditions: in Ref. [6] the condition conservation of heat balance used on a surface, and in our case—a condition of a constant of surface temperature is considered.

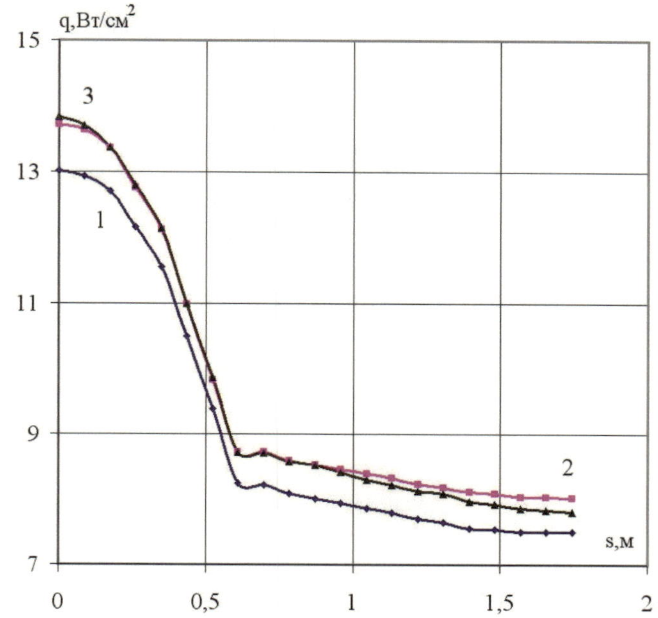

Figure 27. Heat transfer to MSRO vehicle surface in case of non-catalytic wall; 1: equilibrium flow, 2: non-equilibrium flow for three-temperature model [10], 3: non-equilibrium flow for two-temperature model [38]. V_∞ = 5223 m/s, $\rho_\infty = 2.93 \times 10^{-4}$ kg/m^3.

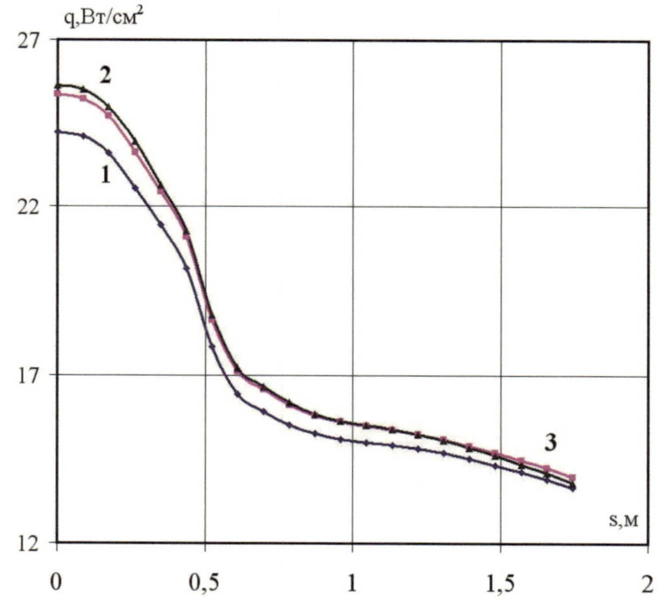

Figure 28. Heat transfer to MSRO vehicle surface in case of ideal catalytic wall; 1: equilibrium flow, 2: non-equilibrium flow for three-temperature model [10], 3: non-equilibrium flow for two-temperature model [38]. V_∞ = 5687 m/s, $\rho_\infty = 3.141 \times 10^{-5}$ kg/m^3.

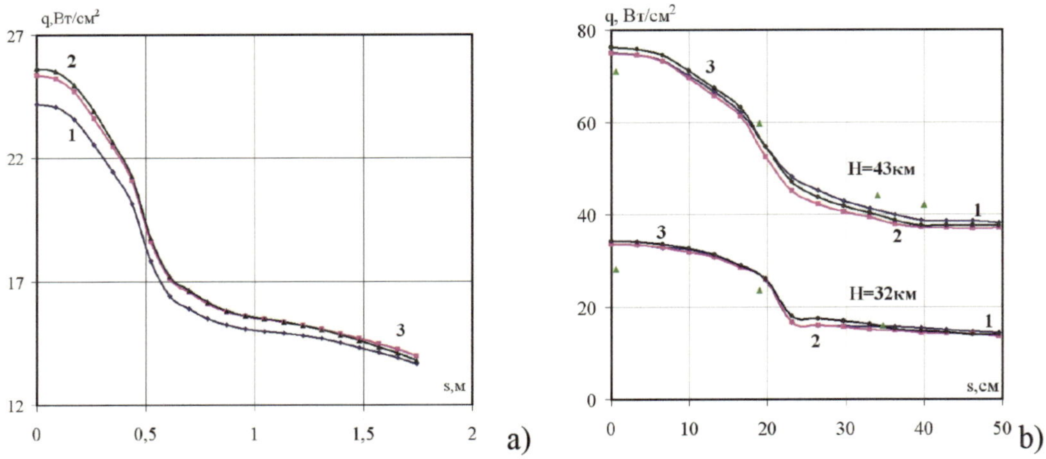

Figure 29. Heat transfer to MARS EXPRESS vehicle surface in case of non-catalytic (a) and ideal catalytic wall (b); 1: model [30], 2: model of [31], 3: model of [32–34], ▲: data of [6].

Catalytic surface have selectively but significant effect on species and reactions on the surfaces. During entry in Martian atmosphere molecule of carbon dioxide (CO_2) dissociated on CO and O. The most part of energy of flow (¾) spends on dissociation. For high catalytic surface kw (surface assist to recombination the atoms to molecules), chemical reactions that spend on dissociation transfer (partially or fully) their energy back to the surface and produce on an additional heating. For low catalytic surface, recombination does not occurs and atoms move downstream and take away the energy of dissociation and additional heating does not happen. It is possible to diminish temperature of surface on 300–500 K when temperature of flow is about 1800–2000 K.

In **Figure 30**, the heat transfer to ideal catalytic and non-catalytic surface of the vehicle for conditions of flow MSRO $V_\infty = 5687$ m/s, $\rho_\infty = 3.14 \times 10^{-5}$ kg/m^3 and $T_w = 1500$ K are displayed.

In **Figure 31**, the heat transfer to non-catalytic wall of the vehicle obtained for conditions of flow V_∞ = 3998 m/s, ρ_∞ = 3.0 × 10^{-4} kg/m^3. In **Figure 31**, the distinction between data of corresponding curves pick up to 15%. In this case, the flow in a shock layer is characterized by smaller temperatures (velocity in an external flow was less) than the regime which correspond to curves of **Figure 30**.

Therefore, it is marked a smaller divergence in values of reaction rate constants and as consequence at values of a heat fluxes for different models.

In **Figure 32**, the heat flux to non-catalytic wall for numbers Sc = 0.45; 0.65 and two conditions of a flow are shown. It is evident that for non-catalytic surfaces the value of Schmidt's number practically does not take an influence on the heat transfer. The heat flux is determined basically just by heat conductivity. In this case near the surface of a body there are not recombination reactions. The products of dissociation are carrying out with themselves the energy of dissociation. As a result of additional heating, the body caused by diffusion processes it does not occur.

In **Figure 33**, the heat transfer to non-catalytic wall is shown (conditions of flow $V_\infty = 5223$ m/s, $\rho_\infty = 2.9 \times 10^{-4}$ kg/m^3) for different values of Prandtl's number: Pr = 0.66 and Pr = 0.75. In a case

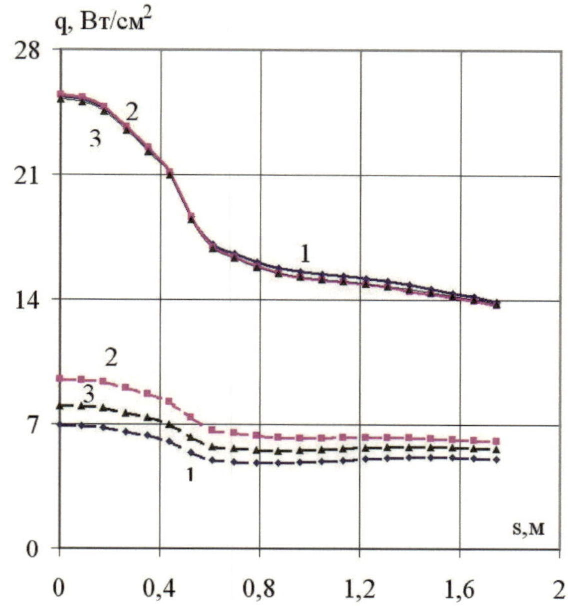

Figure 30. Heat transfer to MSRO vehicle surface in case (——) idealcatalytic wall; (- - - -) non-catalytic wall; $V_\infty = 5687$ m/s, $\rho_\infty = 3.14 \times 10^{-5}$ kg/m^3; 1: model of [30], 2: model of [31], 3: model of [32–34].

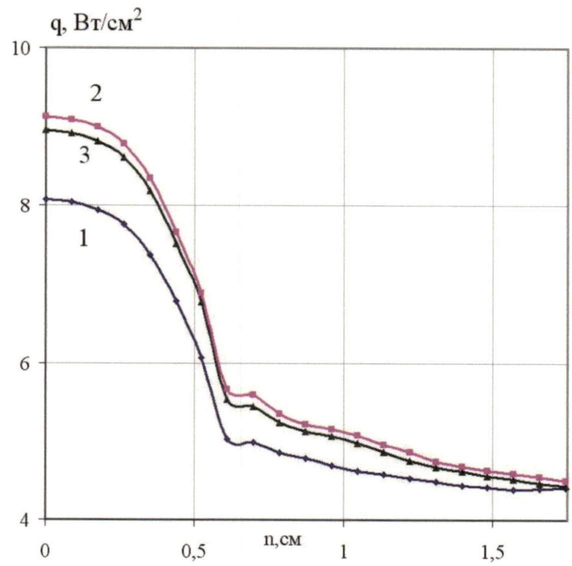

Figure 31. Heat transfer to MSRO vehicle surface; non-catalytic surface $V_\infty = 3998$ m/s, $\rho_\infty = 3.0 \times 10^{-4}$ kg/m^3. 1: model of [30], 2: model of [31], 3: model of [32–34].

of non-catalytic wall, a leading role has another dimensionless factor—Prandtl's number. You can see that the increase in value of Prandtl's number from 0.66 till 0.75 leads to reduction of a heat flux value approximately on 10%.

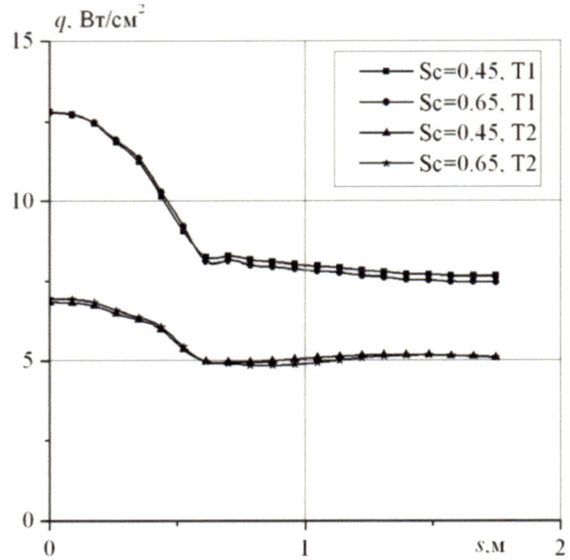

Figure 32. Heat transfer to vehicle surface; T1: V_∞ = 5223 m/s, ρ_∞ = 2.93 × 10^{-4} kg/m^3; T2: V_∞ = 5687 m/s, ρ_∞ = 3.14 × 10^{-5} kg/m^3.

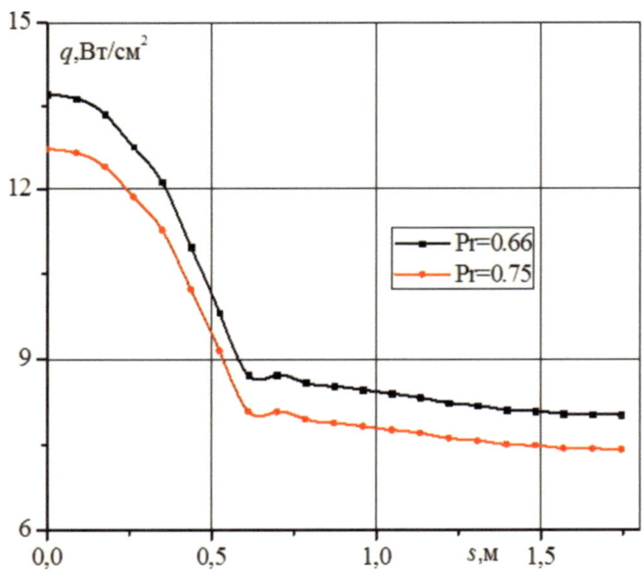

Figure 33. Heat transfer to non-catalytic wall for different Prandtl's number; V_∞ = 5223 m/s, ρ_∞ = 2.9 × 10^{-4} kg/m^3.

7.5. Influence of catalytic wall on a heat transfer

The heat flux to a surface of the vehicle without taking into account radiation effect in one-temperature approach is determined as:

$$\mathbf{q} = -\lambda\nabla T + \sum_{i=1}^{N} h_i \mathbf{J}_i - p\sum_{i=1}^{N} D_{T_i}\mathbf{d}_i \tag{35}$$

where h_i is the enthalpy of ith components of a mixture, λ is the coefficient of heat conductivity of all degrees of freedom which are taking place in a condition of local thermal balance. The second component with in the right part (5) defines the diffusion component of a heat flux and the third term characterizes influence of thermo-diffusion on heat exchange.

As show results of calculations the contribution of the third component on value of a total heat transfer makes less than 1%. It means that influence of thermo-diffusion on heat exchange is small. The mass transfer processes play the important role in definition of a heat transfer to a surface. Really diffusion component for a heat transfer for the considered conditions of a flow makes a significant part (50–75%) from value of a full heat flux. In this connection, the correct account of diffusion processes has great importance.

As already it has been shown that the main factors influencing heat transfer to a surface of the vehicle, it is heterogeneous recombination of component in dependence on catalytic properties of a surface. We try to explain the influence of chemical reactions on parameters of flow in a shock layer and value of a heat transfer for MSRO vehicle in a case of non-catalytic surfaces. And then it is compared with received results with the corresponding data for ideal catalytic surface.

For calculation, diffusion fluxes were used: Fick's law with the adjusting amendment; Fick's law in standard form in which effective diffusion coefficients are calculated through binary diffusion coefficients and concentrations from point of view of the formula; and Fick's law in the usual form in which Schmidt's number was considered equal to all a component and to be constants.

Depending on the manner of representation of diffusion, the heat transfer can differ essentially up to 30%. Let us notice that more correct way of the diffusion definition that is used for the Fick's law with the amendment gives the values of a heat transfer exceeding on 10% corresponding values received from point of view of Fick's law in the standard form. It is evident that the variant with Schmidt's number Sc = 0.45 results most close to corresponding data that it is obtained at a correct way of the diffusion account [38, 39].

For non-catalytic surface, main effect will be play another parameter—Prandtl's number. In **Figure 34**, you can see the heat transfer to non-catalytic surface (V_∞ = 5223 m/s, $\rho_\infty = 2.9 \times 10^{-4}$ kg/m^3) for different Prandtl's number: Pr = 0.66 and Pr = 0.75. Increase of Prandtl's number from 0.66 until 0.75 leads to decreasing heat transfer on 10%.

In **Figure 35**, the heat flux for two types of a surface and two flow conditions is shown. You can see that value of a heat transfer to non-catalytic surfaces is approximately 3–4 times less than corresponding value of a heat transfer in a case ideal catalytic surface.

Figure 36 shows values of a heat flux to a surface having various catalytic property. The curve 2 (a variant of a surface with final catalytic) is obtained for the following parameters:

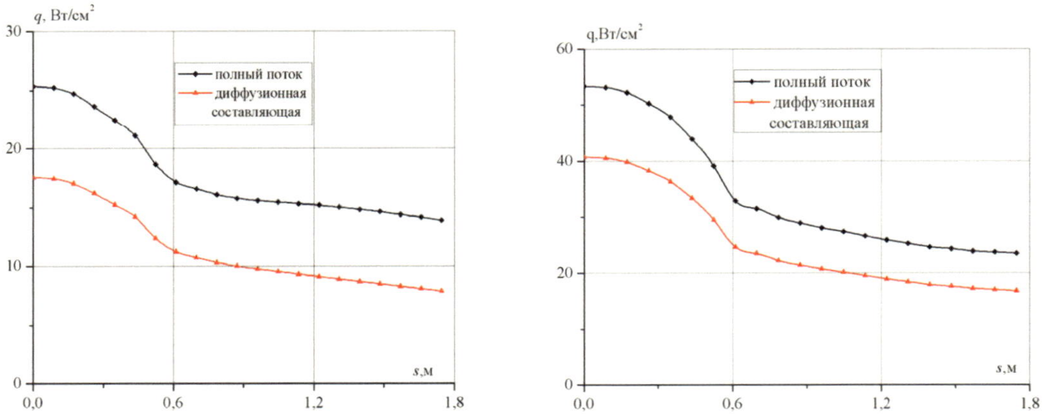

Figure 34. Heat transfer to vehicle surface. Black line—total heat flux; red line—diffusion part; (a) V_∞ = 5687 m/s, (b) V_∞ = 5223 m/s.

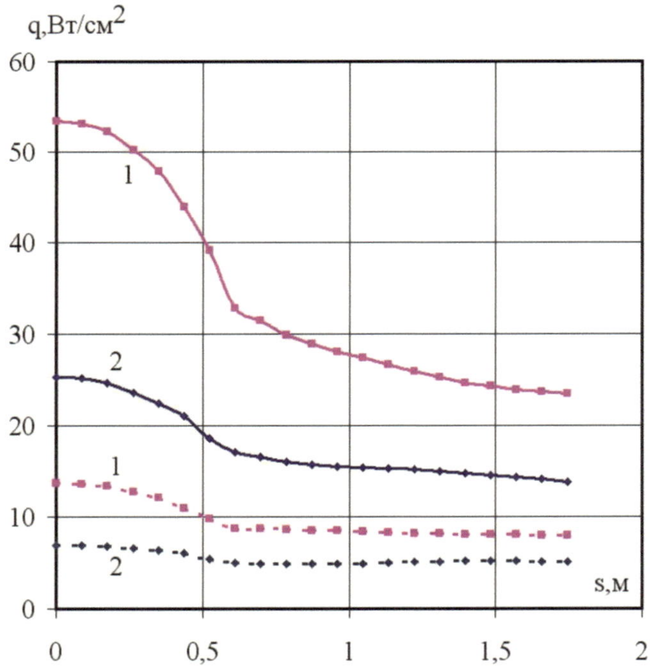

Figure 35. Heat transfer to MSRO vehicle surface. 1: V_∞ = 5223 m/s, H = 40 km; 2: V_∞ = 5687 m/s, H = 60 km.

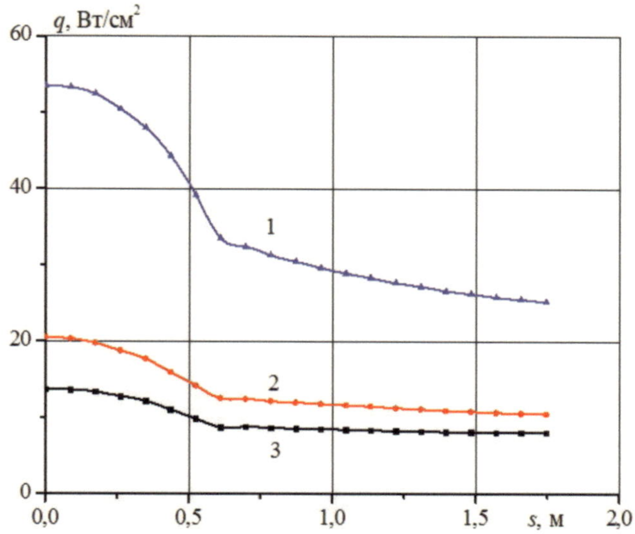

Figure 36. Heat transfer to MSRO vehicle surface. 1: Ideal catalytic wall, 2: finite catalytic, k_{wco} = 0.77 m/s, k_{wo} = 1 m/s, 3: non-catalytic wall, V_∞ = 5223м/c, H = 40 km.

probability of recombination reactions—γ_w = 2.7 × 10^{-3}; catalytic constants—k_{wo} = 1 m/s, k_{wco} = 0.77 m/s (see the formula 30). From the presented data, you can see that due to use low catalytic coverings it is possible to lower a heat transfer to a surface of the vehicle on a significant part of a trajectory in some times.

The submitted data show an insignificant influence of processes of vibration relaxation on a heat transfer to a surface of MSRO vehicle. It is also possible to note that taking into account of complex internal structure of molecules CO_2 and exchanges of vibration energy between modes does not take an influence on a heat transfer to a surface. It testifies to legitimacy of application of the simplified models of vibration kinetics at the solution of the given class of problems. The fact of weak influence of a vibration relaxation on a heat transfer to a surface of the vehicle is possible to explain for considered conditions of a flow and the considered form of a surface intensive process dissociation molecules of carbon dioxide gas in a shock layer is observed, and also by fast process of an vibration relaxation of molecules CO_2 as a result of which thermodynamic equilibrium is present almost in all shock layer.

We would like to consider the influence of the form of the blunted body on parameters of a non-equilibrium flow in modeling an atmosphere of Mars a flow of carbon dioxide gas. Let us consider the heat transfer along surface of spherically blunted cones with various half opening angles.

In a vicinity of the stagnation point on a spherical part of cone surface with various angle of opening, the solution practically coincides. On a conic surface, the size of parameters of a flow essentially depends on an angle of opening a cone that affects on value of a heat flux to a surface (**Figure 37**).

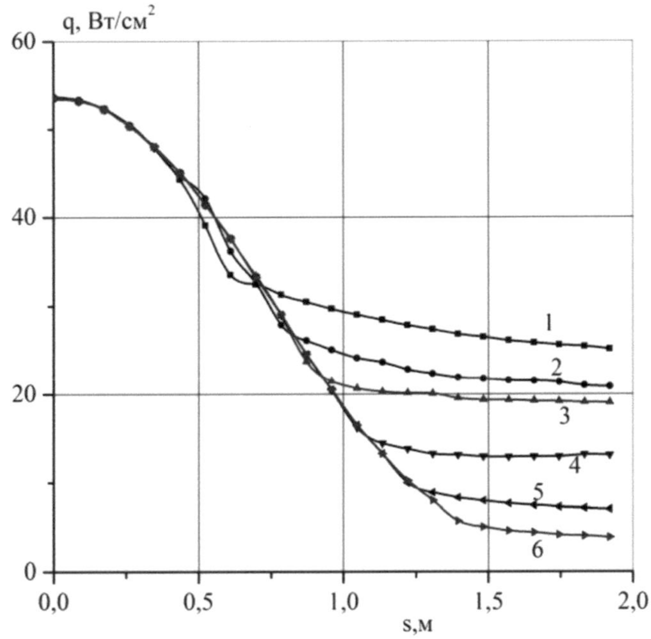

Figure 37. Heat transfer to the surface of blunted cones with different opening angle θ for Mars atmosphere; 1: $\theta = 60°$, 2: 50°, 3: 40°, 4: 30°, 5: 20°, 6: 10°. $V_\infty = 5223$ m/s, $\rho_\infty = 2.93 \times 10^{-4}$ kg/m^3.

As a result of the carried out researches, it is received (calculations were carried out with help of two-temperature model of a vibration relaxation) that for conditions of flows in an atmosphere of Mars corresponding to velocities 4–6 km/s and to altitude H <60 km flight, the vibration relaxation of molecules on a spherical part of a body occurs quickly and thermodynamic equilibrium is present almost in all cross section of a shock layer.

As show numerical calculations for altitude of flight H ≤ 60 km the blunted bodies in an atmosphere of Mars the increase of a heat transfer to a surface taking in account the mechanism of an vibration relaxation makes no more than 20%.

7.6. The role of non-equilibrium radiation

The role of radiation is rather insignificant for descent space vehicle in an atmosphere of Mars with the characteristic sizes 1–2 m and at entrance velocity of $U_\infty \approx 5 - 6$ km/s. On vehicles which sizes will be at 5–10 times more, the radiant flux can be comparable with convective. The pilot and automatic expeditions with use of diving is perspective due to aerodynamic braking devices in the top layers of Mars atmosphere with U 6–8 km/s up to 3.3 km/s with the subsequent exit into basic orbits around of Mars. In this case the radiant flux is determined by non-equilibrium radiation as the vehicle penetrates an atmosphere of Mars at heights $H > 30$ km in which physical and chemical processes in a shock layer is essentially non-equilibrium. At $H < 30$ km, the radiation flux to space vehicle is determined by equilibrium radiation and its level is insignificant. At hyperbolic velocities of an entrance of flight, the radiation flux can be strong that play an essential role at a stage of a choice of the of descent trajectory and heat protection system of space vehicle.

Thermo-chemical model of the CO_2-N_2 mixtures for the calculation of the non-equilibrium ultra-violet (UV) molecular band radiation in the high-temperature shock layer around the capsule during an entry is considered [19].

In the frontal part of a thin shock layer radiation, absorption is small enough. Thus, the gas is assumed to be transparent to radiation in the relaxation zone of the shock layer. Radiation intensity is calculated in the approximation of volume luminescence. Radiation processes involving excited particles are considered as spontaneous radiating transitions, excitation and deactivation of the electronic states of the molecules, impacts of electron and heavy particles, etc. [40, 41]. Corresponding equation of excitation and deactivation of the electronic states determine a concentration of each component. Flow parameters across the shock layer are calculated and the spectral structure of radiation is obtained. For the analysis of the non-equilibrium radiation, the results of some theoretical and experimental studies behind shock waves in CO_2-N_2 mixtures used. In the relaxation zone of the vehicle shock layer, the Boltzmann distribution of the electronically excited states of atoms and molecules does not exist under the considered Martian entry conditions. It leads to a significant deviation of the radiation intensity from local equilibrium especially for low gas density.

The suggested thermo-chemical model of the $CO_2 + N_2$ mixture contains 10 neutral chemical species: CO_2, CO, CN, NO, N_2, O_2, C_2, C, N, O, 4 molecular ions and free electrons: CO^+, NO^+, N_2^+, O_2^+, e^-, and 12 electronically excited states of diatomic molecules: $CO(A'\Pi)$, $CO(b^3\Sigma^+)$, $CN(B^2\Sigma^+)$, $CN(A^2\Pi)$,$NO(B^2\Pi)$, $NO(A^2\Sigma^+)$, $NO(C^2\Pi)$, $NO(D^2\Sigma^+)$, $N_2(A^3\Sigma_u^+)$, $N_2(B^3\Pi_g)$, $O_2(B^2\Sigma_u^-)$, $C_2(d^3\Pi_g)$.

The thermo-physical properties of chemical species are taken from Ref. [42]. Thus, 19 chemical reactions and 33 reactions of the excitation of the electronic states of molecules are taken into account. The reactions with the participation of the neutral and charged particles in a high-temperature Martian atmosphere are considered. The rate constants of the basic chemical reactions have been collected from the literature [29, 31–34, 43].

Practically, it is convenient to use simplified radiation models that are capable to estimate radiation emission with sufficient accuracy. The estimations have shown that the gas is transparent to UV molecular radiation in the shock layer under the considered conditions. And it is possible to calculate radiation intensity with the approximation of volume luminescence. The "just overlapping line model" model is used to calculate spectral distribution of non-equilibrium molecular band radiation. The model considers a spectrum consisting of only one branch of rotational lines. The shock layer is optically thin for spectral range considered so the process of light absorption is not taken into account. It is shown that the depletion of electronically excited states of molecules due to spontaneous radiation transitions has a great effect on excited state populations and must be necessarily accounted for under the MSRO trajectory conditions. Spontaneous radiation emission leads to violation of Boltzmann approximation for excited state populations. The molecular band radiation results obtained in the local equilibrium approximation (i.e. the supposition of the Boltzmann distribution of molecules on electronic states) strongly overestimates non-equilibrium radiation values and cannot be used even for preliminary predictions.

Calculations of the convective heat flux and the non-equilibrium radiation were carried out for the MSRO vehicle entering into the Martian atmosphere. The wall is assumed non-catalytic. Trajectory parameters are presented in **Table 2**.

In **Figure 38a**, distributions of temperature along the stagnation line are shown. In all cases, it follows from a result the model viscous shock layer is realized. Boundary layer thickness takes approximately 1/4 from the shock layer thickness.

The main contribution to a radiation in the shock layer is produced by the bands of the molecular systems as found from theory and experiments. Main source of the shock layer radiation are the molecules that form as a result of chemical reactions.

Molecules CO, O_2, CN, C_2 are formed only as a result of the chemical reactions and the information about vibration states of these molecules are absent. For these molecules, their vibration modes are in thermal equilibrium. Besides, there is a significant amount of oxygen atoms due to the fast dissociation of the molecules CO_2 behind shock wave. Oxygen atoms have large enough cross sections for V-T processes energy exchange. In the free stream, CO_2 and N_2 molecules have almost zero vibration energy, therefore, for them in a shock layer there is an area with non-equilibrium vibration.

V, m/s	T_∞, K	ρ_∞, kg/m^3
3536	140	2.82×10^{-5}
3998	140	3.07×10^{-4}
5223	140	2.93×10^{-4}
5687	140	3.125×10^{-5}

Table 2. Trajectory parameters.

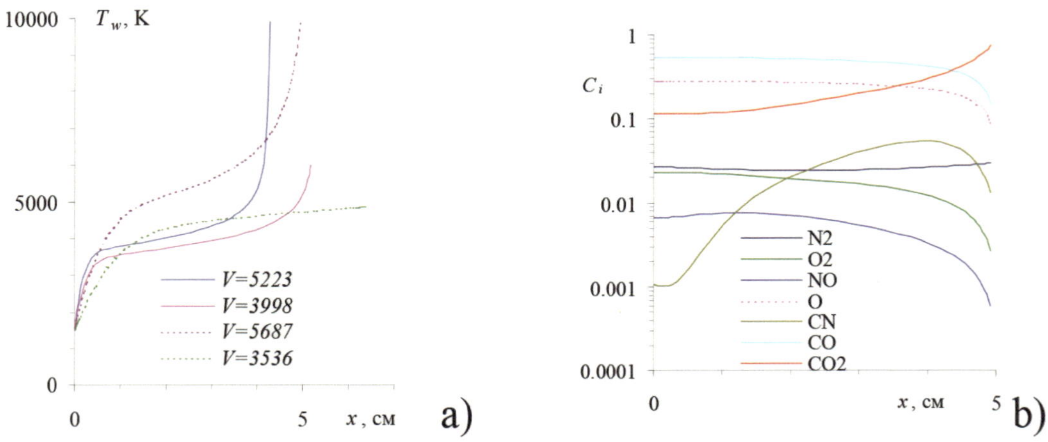

Figure 38. Translational temperature (a) and mass concentrations, $V = 5687$ m/s, (b) along stagnation line.

The radiation intensities are obtained for the 11 strongest systems of molecular bands: 3rd and 4th positive band of CO molecule, the red and violet band of CN molecule, the Schuman-Runge system of O_2 molecule, the β, γ, δ, ε systems NO molecule, the Swan band of the C_2 molecule, 1st positive band of N_2 molecule.

In **Figure 38b**, distributions of mass concentrations of neutral chemical species along the stagnation line are shown for 5687 m/s. Note a maximum of concentration CN in the relaxation zone behind the shock wave. With increasing velocity, the degree of dissociation of the molecules CO_2 increases and concentration of CN molecules become larger.

In **Figure 39**, distributions of volume concentration of the electronically excited states along the stagnation line are shown. Continuous lines show the values obtained with non-equilibrium approach, while dotted lines correspond to the local equilibrium approximation. Practically, all the excited levels, except O_2(B), reach a maximum near the shock wave. Such a fact is due to the larger values of the temperature in this region. Populations of the molecules containing carbon, except CN (A) and CO (d3) are essentially smaller (approximately in 100 times), the corresponding Boltzmann distributions. Thus, in the relaxation zone of the shock layer, the Boltzmann distribution of atoms and molecules on the electronically excited states is violated. It leads to a significant deviation of the radiation intensity from that corresponding to local equilibrium case especially for low gas density.

For the molecules forming behind the shock waves, the populations of NO (D), NO (C) states are close to equilibrium. The populations of the rest electronic states differ from their equilibrium values but in a less degree than for the molecules containing carbon.

Intensity of radiation strongly decreases near the body surface. Thus for rather low value of temperature across the boundary layer, it does not bring an appreciable contribution to the radiation heat transfer. Boundary conditions on the surface of the body, in particular, the catalytic condition does not affect value of radiation heat transfer.

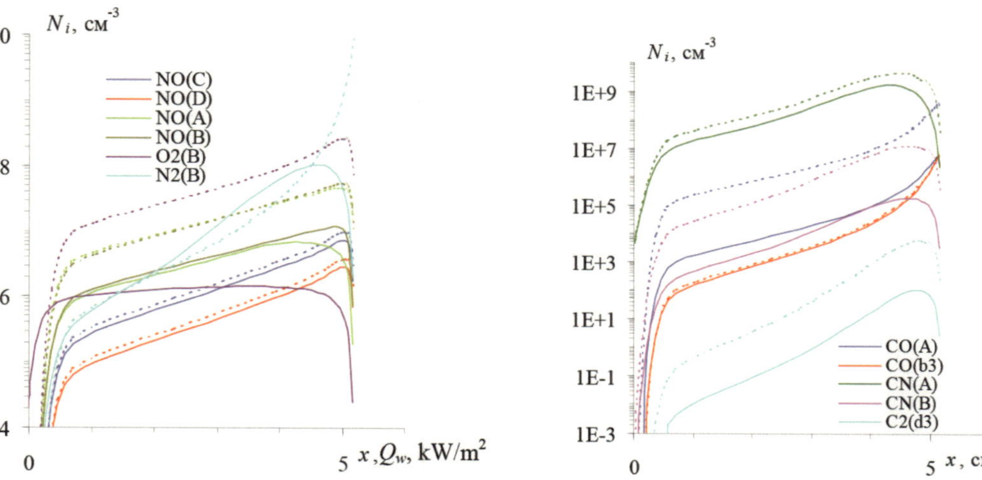

Figure 39. Populations of electronically excited states along stagnation line, V = 3998 m/s.

Spectral distributions of radiation intensity from ultra-violet up to near infra-red are estimated. The data of non-equilibrium radiating processes and the local equilibrium approach are compared. Use of the local equilibrium modifies drastically the spectral distribution of radiation intensity.

In **Figure 40**, the convective heat fluxes for different trajectory points are shown. It is possible to divide the distribution of the convective heat flux along the surface in three distinct regions. The maximum heat transfer occurs at the stagnation point. Then, along the spherical part, the heat transfer decreases as the pressure drops. The heat transfer along the conic part is almost constant. At last, there is a local increase of the heat transfer in the shoulder region connected with an increase of the velocity gradient.

In **Figure 41**, convective and radiation heat transfer values are compared for different trajectory points. Convective heat flux is predominant compared to the non-equilibrium radiation flux. With account of non-equilibrium character of collision-radiation processes in the shock layer values of radiation heat transfer are several orders (from 10^2 to 10^7 times) below convective ones for the considered trajectory points. For a correct prediction of heat transfer and surface temperatures near space vehicle at entry conditions in the Martian atmosphere, the careful examination of theoretical and experimental catalytic properties results of a of thermal protection covering are required.

However, for the local equilibrium approach, the radiation flux is close to the convective value only for the trajectory point (V = 5687 km/s, ρ_∞ = 3.125 × 10^{-5} kg/m^3). The non-equilibrium radiation flux is one order of magnitude smaller less than the flux obtained under the local equilibrium assumption.

To assess the influence of the catalytic wall on heat transfer and radiation equilibrium temperature of the surface thermal protection calculations for 120 degree cone blunted on sphere of radius R = 0.7 m at speed V = 6150 km/s considering a pure CO_2 atmosphere at an altitude of 40 km have been carried out. Calculations were made for four values of the recombination

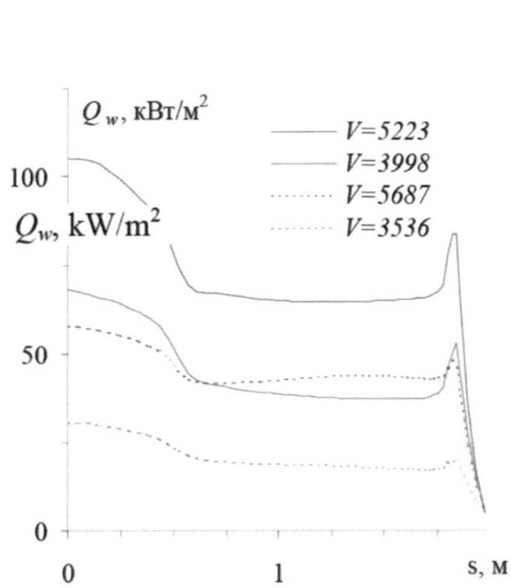

Figure 40. Convective heat flux distributions.

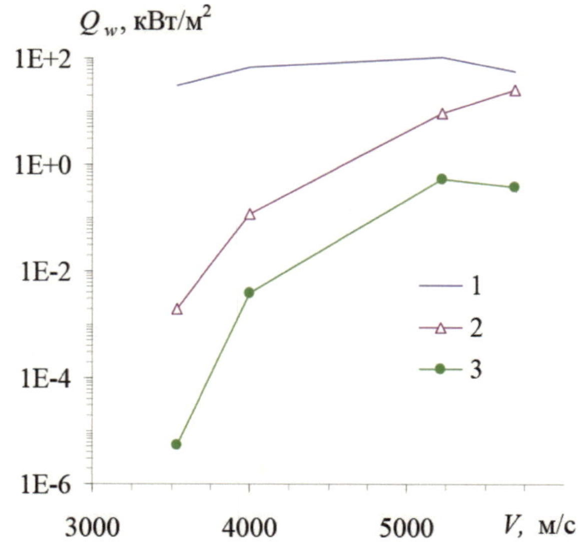

Figure 41. Heat flux at the stagnation point along trajectory, 1: convective, 2: local equilibrium radiation, 3: non-equilibrium radiation.

γ	K_w, m/s	Q_w, kW/m^2	T_w, K
10^{-4}	0.035	281	1530
10^{-3}	0.37	380	1650
10^{-2}	4	692	1915
10^{-1}	40	897	2045

Table 3

probability that are typical for heat-shielding materials of different types. Results of calculations are presented in **Table 3**.

According to the results, the ratio between the lowest heat flux and largest one is close to three. The equilibrium surface temperatures when blackness of a surface ε = 0.9 can differ more than 500 K. For a correct prediction of heat transfer and surface temperatures, careful experimental researches about the catalytic properties of the thermal protection covering are required.

8. Conclusions

Non-equilibrium flows of the reacting mixture $CO_2/CO/O_2/C/O$ in a viscous shock layer near a spacecraft entering the Mars atmosphere are studied using the accurate three-temperature model developed on the basis of the kinetic theory methods. Gas dynamic parameters, transport coefficients in a shock layer, and heat fluxes to the body surface are calculated for non-catalytic and fully catalytic surfaces. The results are compared with the ones obtained in the simplified two-temperature approximation and in the one-temperature approach for weak deviations from thermal equilibrium. A considerable influence of CO_2 vibration excitation on the flow parameters and transport properties in a shock layer is found. The difference between the results obtained using the accurate and simplified vibration non-equilibrium models are weak under conditions considered in the paper. This justifies the validity of the approximate two-temperature model under the re-entry conditions. It is shown that difference in reaction rate constants practically has small influence on value of a heat transfer to ideal catalytic surfaces of the vehicle. In a case of non-catalytic surfaces difference in value of the heat transfer obtained by different models can be essential up to 30%. The effect of bulk viscosity in a shock layer is studied. Including this coefficient to the fluid dynamics equations improves the accuracy of the heat flux calculation up to 10%.

The heat flux to ideal catalytic surface for the considered conditions of a flow can up to four times surpasses a heat transfer to non-catalytic wall.

The influence of the different chemical reactions models: (1) Mc. Kenzie and Arnold chemistry model, (2) Park's model, amd (3) model of S. Losev and others, on component concentrations and heat flux are presented. Numerical calculations of the coefficients of viscosity and heat conductivity give the close result for the heat flux for different models comparing with results obtained by exact kinetic theory. The diffusion parameters affects on the magnitude of the heat flux especially in the case of a catalytic wall. The pressure- and thermo-diffusion influence on heat flux are small. The different models of the vibration relaxation of CO_2 considered. They give approximately the same values of main properties of the flow. The catalytic properties of the surface are most important for a valid determination of the heat flux to the wall. The insertion of bulk viscosity into the equations leads to the small increase of the heat fluxes.

Author details

Yuriy D. Shevelev

Address all correspondence to: shevelev@icad.org.ru

The Institute for Computer Aided Design of RAS, Moscow, Russia

References

[1] Kay RD, Netterfield MP. Thermochemical Non-Equilibrium Computation For A Mars Entry Vehicle. AIAA Paper. No 93-2841; 1993

[2] Gupta RN, Lee KP. An Aerothermal Study of MESUR Pathfinder Aeroshell. AIAA Paper. No 94-2025; 1994

[3] Gallis MA, Harvey JK. Analysis of Non-Equilibrium in Mars Atmosphere Entry Flows. AIAA Paper. No 95-2095; 1995

[4] Problems of Aerothermoballistics, Radiation Gasdynamics, Heat and Mass Transfer for Planet Sample Return Missions. Project Technical Report of ISTC No 1549-00; 2003. http://jan.imec.msu.ru/~avogadro/Mars/Mars1549/index.htm

[5] Rouzaud O, Zhlukhtov S, Egorov I, Fletcher D, Gromov V, Nagnibeda E, Shevelev Yu, Omaly P. Numerical, analytical and experimental investigation of convective and radiative heating of a martian descent module. European Space Agency, ESA SP (583); 2005. pp. 93-98

[6] Afonina NE, Gromov VG. Numerical Modeling of Martian Descent Module Flow. Institute of Mechanics in Moscow State University. Aeromechanics and Gas Dynamics: Moscow. 2001;**2**:35-47

[7] Nagnibeda E, Kustova E. Kinetic Theory of Transport and Relaxation Processes in Non-equilibrium Reacting Gas Flows. Saint Petersburg: Saint Petersburg University Press; 2003

[8] Kustova E, Nagnibeda E. On a correct description of a multi-temperature dissociating CO_2 flow. Chemical Physics. 2006;**321**:293-310

[9] Kustova EV, Nagnibeda EA, Shevelev YD, Syzranova NG. The influence of CO_2 kinetics on the hypersonic flow near blunt bodies. American Institute of Physics, AIP Conference Proceedings. 2012;**1501**:1102-1109

[10] Shevelev YD, Syzranova NG, Kustova EV, Nagnibeda Numerical EA. Investigation of hypersonic fluid discending flow in Mars atmosphere. Mathematical Models and Computer Simulations. 2011;**3**(2):205-224

[11] Kustova EV, Nagnibeda EA, Shevelev YD, Syzranova NG. Non-equilibrium kinetics and transport processes in a hypersonic flow of CO_2/CO/O_2/C/O mixture. American Institute of Physics, AIP Conference Proceedings. 2011;**1333**(Part 1):1227-1232

[12] Kustova EV, Nagnibeda EA, Shevelev YD, Syzranova NG. Comparison of different models for non-equilibrium CO_2 flows in a shock layer near a blunt body. Shock Waves. 2011;**21**(3): 273-287

[13] Shevelev YD, Syzranova NG. The effect of multi-component diffusion on supersonic flow of carbon dioxide past bodies. High Temperature. 2009;**47**(6):829-835

[14] Kustova EV, Nagnibeda EA, Shevelev YD, Syzranova NG. Comparison of non-equilibrium supersonic CO_2 flows with real gas effects near a blunt body. European Space Agency, ESA SP. 2009;659

[15] Shevelev YD. Three-Dimensional Problems of Computational Fluid Dynamics. Moscow: Nauka; 1986

[16] Golovachev YP. Numerical Modeling of Viscous Flows in Shock Layer. Moscow: Nauka; 1996

[17] Kustova EV, Nagnibeda EA, Shevelev YD, Syzranova NG. Non-equilibrium supersonic CO_2 flows with real gas effects near a blunt body. AIP Conference Proceedings. 2009). American Institute of Physics; **1084**:831-836

[18] Shevelev YD, Syzranova NG, Kustova EV, Nagnibeda EA. Bulk-viscosity effect on CO_2 hypersonic flow around blunt bodies. Doklady Physics, Pleiades Publishing, Ltd. 2015; **60**(5):207-209

[19] Gorshkov AB, Vlasov VI, Zalogin GN, Shevelev YD. Numerical Investigation of Non-Equilibrium Radiation of CO_2-N_2 Mixtures in a Shock Viscous Layer; 2009. European Space Agency, ESA SP.659 SP

[20] Marrone PV, Treanor CE. Chemical relaxation with preferential dissociation from excited vibrational levels. Physics of Fluids. 1963;**6**(9):1215

[21] Wilke CR. Diffusional properties of multicomponent gases. Journal of Chemical Engineering Progress. 1950;**46**(2):95-104

[22] Mason EA, Saxena SC. Approximate formula for the conductivity of gas mixture. Physics of Fluids. 1958;**1**(5):361-369

[23] Blottner FG, Curties CF, Bird RB. Chemically Reacting Viscous Flow Program for Multi-component Gas Mixtures. Report No. SC-RR-70-751. Albuquerque New Mexico: Sandia Laboratories; Dec. 1971

[24] Armaly B, Sutton K. Thermal Conductivity of Partially Ionized Gas Mixtures. AIAA Paper. 1982. No. 469. 6pp

[25] Scott CD. Reacting shock layers with slip and catalytic boundary conditions. AIAA Journal. 1975;**13**(10):1271-1278

[26] Jameson A, Yoon S. A LU-SSOR scheme for the Euler and Navier-Stokes equations. AIAA Paper; 1987. No 87-0600. 11p

[27] Gnoffo A, Wielmuenster KJ, Hamilton HH. Computational Aerothermodynamic Design Issues for Hypersonic Vehicles. AIAA-Paper, 97-2473. pp. 1–35

[28] Kumar A, Jain AC. Nonequlibrium merged stagnation shock layers at hypersonic speeds. International Journal of Heat and Mass Transfer. 1975;**18**(10):1113-1118

[29] Lunev VV. Flows of Real Gas at High Velocities. Moscow: Fizmatlit; 2007. 760p. (in Russian)

[30] McKenzie RL, Arnold JO. Experimental and theoretical investigations of the chemical kinetics and nonequilibrium CN radiation behind shock waves in CO_2-N_2 Mixture. AIAA Paper. No 67-322; 1967

[31] Park C, Howe JT, Jaffe RL, Candler GV. Review of chemical-kinetic problems of future NASA missions, II: Mars entries. Journal of Thermophysics and Heat Transfer. 1994;**8**(1): 9-23

[32] Kovach EA, Losev SA, Sergievskaya AL. Models of two-temperature chemical kinetics for description of dissociation in strong shock waves. Chemical Physics. 1995;**14**(9):1353-1387 (in Russian)

[33] Ibragimova LB, Smehov GD, Shatalov OP. Constants of dissociation reaction rates of two-atom molecules in thermal equilibrium conditions. Izv. Ross. Akad. Nauk, Mekh. Zhidk. Gaza. 1999;(1):181-186

[34] Ibragimova LB. Constants of chemical reactions rates for high temperature gas CO_2. Mathematical Modeling. 2000;**12**(4):3-19. (in Russian)

[35] Park C. Nonequilibrium Hypersonic Aerothermodynamics. N.Y.: Wiley-Intern Publ; 1990

[36] Kustova EV, Nagnibeda EA, Shevelev YD, Syzranova NG. Non-equilibrium kinetics and transport processes for hypersonic flows of carbon dioxide gas. Physical-Chemical Kinetics in gas Dynamics. 2008;**6**. http://www.chemphys.edu.ru/media/files/2008-04-04-001.pdf. (in Russian)

[37] Taylor RL, Bitterman S. Survey of vibrational relaxation data for processes important in the CO_2+Ar laser systems. Reviews of Modern Physics. 1969;**41**(1):26-47

[38] Shevelev YD, Syzranova NG. The influence of different chemical kinetics models on supersonic carbon dioxide gas flows past blunt bodies. Physical-Chemical Kinetics in Gas Dynamics. 2007;**5**. http://www.chemphys.edu.ru/media/files/2007-12-17-001.pdf. (in Russian)

[39] Shevelev YD, Syzranova NG. Modeling of diffusion in multicomponent gas media. Physical-Chemical kinetics in Gas Dynamics. 2011;**12**. URL:http://www.chemphys.edu.ru/pdf/2011-05-26-001.pdf. (in Russian)

[40] Plastinin YA, Vlasov VI, Gorshkov AB, Kovalev RV, Kuznetsova LA. Analysis of Non-Equilibrium Radiation for Low Density Hypersonic Flows at Low to Moderate Velocities. AIAA Paper; 1998, No 98-2466

[41] Kamenshchikov VA, Plastinin YA, Nikolaev VM, Novitskiy LA. Radiation Properties of Gases at High Temperatures. Moscow: Mashinostroenie; 1971 (in Russian)

[42] Glushko VP, editor. Thermodynamic Tables of Chemical Species. Moscow; 1978 V.2 (in Russian)

[43] Losev SA, Kozlov PV, Kuznetsova LA, Makarov VN, Romanenko YV, Surzhikov ST, Zalogin GN. Radiation of Mixture CO_2-N_2-Ar in Shock Wave: Experiment & Modeling. Third European Symposium on Aerothermodynamics for Space Vehicles. Noordwijk: ESTEC; 1998

Chapter 4

Launcher Aerodynamics: A Suitable Investigation Approach at Phase-A Design Level

Giuseppe Pezzella and Antonio Viviani

Additional information is available at the end of the chapter

http://dx.doi.org/10.5772/intechopen.70757

Abstract

This chapter deals with launcher aerodynamic design activities at phase-A level. The goal is to address the preliminary aerodynamic database of a typical launch vehicle configuration as input for launcher performances evaluations, control, sizing, and staging design activities. In this framework, different design approaches relying on both engineering and numerical methods are considered. Indeed, engineering-based aerodynamic analyses by means of a three-dimensional panel methods code, based on local surface inclination theory, were performed. Then, accuracy of design analysis increased using steady-state computational fluid dynamics with both Euler and Navier-Stokes approximations.

Keywords: launcher vehicles, aerodynamic design, subsonic, transonic, supersonic and hypersonic speed flows, computational fluid dynamics, panel methods aerodynamics

1. Introduction

During the design phase of launchers, the aerodynamic characterization represents a fundamental contribution. Usually, it is accomplished by means a hybrid approach encompassing wind tunnel testing (WTT) and computational fluid dynamics (CFD) investigations [1]. This combined design approach (i.e., WTT and CFD analyses) is extremely reliable in providing high quality data as input for launchers' sizing, performance evaluations, control, and staging dynamics [2]. Indeed, launcher aerodynamics focuses on the assessment of the pressure and skin friction loads the atmosphere determines over the vehicle surface [3]. As well known, these loads result in a global aerodynamic force that acts at the aeroshape center of pressure (CoP) which generally does not coincide with the vehicle center of gravity (CoG) [4]. As a result, the related aerodynamic moment acting at the CoG can lead to a stable or unstable behavior of the launcher to account for in the control software [5]. Moreover, the analysis of the flowfield past the launcher is also fundamental to address the effects of aeroshape's structures and protrusions. Indeed, aeroshell steps and gaps determine local pressure (and convective heat flux) overshoots all along the ascent trajectory [6]. This assessment is fundamental for launcher sizing and thermal protection design activities [7].

With this in mind, the present research effort describes typical aerodynamic analyses performed at Phase-A design level [8]. Indeed, engineering-based analyses are carried out by exploiting local surface inclinations methods. After that, fully three-dimensional steady-state CFD analyses have been addressed to feed launcher aerodynamic design in the range between Mach 0.5 and 5.

Nevertheless, this chapter opens focusing attention on the assessment of the reliability of the present numerical design approach. Indeed, a CFD validation study was undertaken in order to highlight the capability of this CFD approach in assessing some critical aerothermal design issues, namely shock-shock interaction (SSI) and shock wave boundary layer interaction (SWIBLI), of vehicle aeroshapes flying at hypersonic speed, like launchers.

Finally, note that numerical flowfield analyses are performed with FLUENT code and perfect gas flow model.

2. CFD validation study

In the last years, CFD has played an important role in hypersonics being able to address particular design issues, such as the well-known SSI and SWIBLI [4]. These flowfield features occur whenever different shocks interact each other or with the boundary layer when a shock impinges on a wall, respectively. For launchers, SSI and SWIBLI phenomena typically take place in the flowfield region within launcher main body and boosters, as shown in the schematics of **Figure 1**.

In this figure, the fairings bow shock meets that of booster, thus originating a SSI. This interaction results in more or less complex shock patterns including shear-layers or jets, which

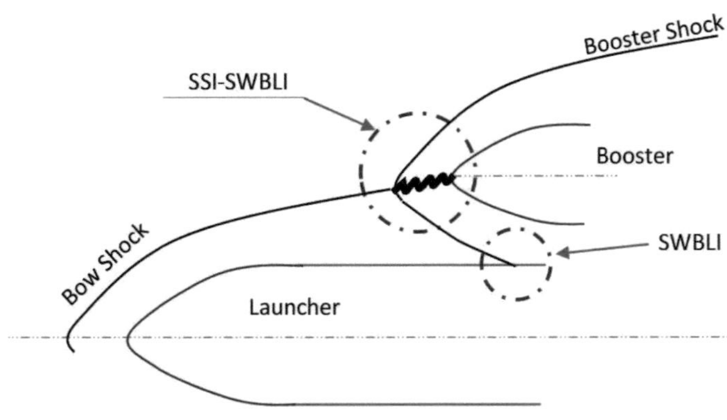

Figure 1. Example of SSI and SWBLI for launchers.

can impinge on the launcher aeroshape and cause local pressure and heat flux overshoots, well in excess of those occurring at stagnation points.

On the other hand, SWBLI occurs, for instance, when the shock resulted from the SSI meets the launcher wall, thus promoting boundary layer separation and transition.

As a result, SSI and SWBLI demand accurate prediction for a reliable and affordable aerothermal design of launcher vehicles.

In this framework, the results of the computational analysis of the flowfield past a double wedge test bed are reported and discussed in detail. This configuration, in fact, is a benchmark as it presents unique flow patterns typical of SSI and SWIBLI. In particular, the experiment of Swantek and Austin was selected and numerically rebuilt [9]. The test bed geometry is shown in **Figure 2**. It is a double wedge with θ_1 = 30° and θ_2 = 55° where the lengths of the first and second face are L_1 = 50.8 mm and L_2 = 25.4 mm, respectively.

Along with the center of the model 19 coaxial thermocouple gauges at 16 different streamwise locations are mounted. Therefore, several experimental data exist for numerical-to-experimental comparisons. The test campaign was performed by using high enthalpy air at the free-stream conditions summarized in **Table 1**.

The numerical rebuilding was carried out by means of a steady-state two-dimensional Reynolds-averaged Navier-Stokes (RANS) simulation performed with the commercial CFD tool Fluent. Air was modeled with a five species chemistry mixture (N_2, N, O_2, O, NO) in

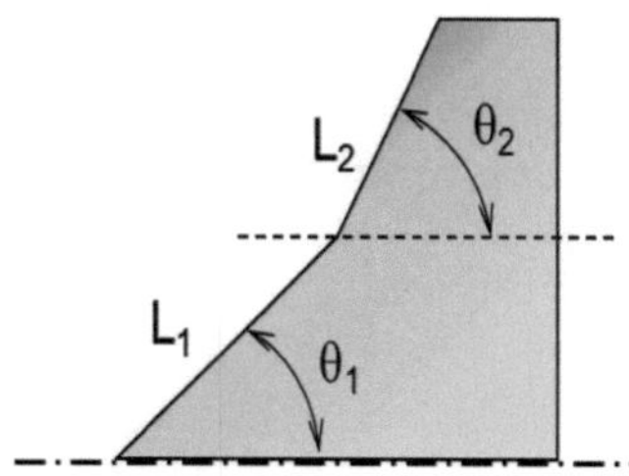

Figure 2. Test bed configuration with quotes.

Parameter	M7_8
Stagnation enthalpy (MJ/kg)	8.0
Mach	7.14
Static temperature (K)	710
Static pressure (kPa)	0.780
Velocity (m/s)	3812
Density (kg/m^3)	0.0038
Unit Reynolds number (10^6/m)	0.435

Table 1. Free-stream conditions of experiment.

thermo-chemical non-equilibrium conditions. Turbulence has been taken into account with the k−ω SST model. The wall was assumed isothermal (T_w = 298 K) and noncatalytic; while, in order to take into account the effects of the boundary layer transition, a trade-off analysis was undertaken (and not shown here for simplicity) in order to determine a proper flow transition location (x_{tr}) to fix along with the first ramp. Results highlighted that x_{tr} = 58% L_1 is a viable option. Further details about the numerical setting can be found in Ref. [10].

A structured multiblock mesh of 433 × 707 points was considered to solve for complex flow structure past the test bed. In particular, a great deal of care was taken in grid development. In fact, the distribution of grid points has been dictated by the level of resolution desired in various areas of the computational domain such as SSI, triple points, shear layer and recirculation region. An example of the computational grid is provided in **Figure 3**.

As far as numerical results are concerned, **Figure 4** shows the qualitative comparison between experimental data (i.e., Schlieren image) and the Mach isolines.

As one can see, CFD results compare rather well with the Schlieren. Indeed, the numerical flowfield presents the same structure as pointed out by the experimental data, as the triple point, due to a strong shock that originates ahead the recirculation bubble, the reattachment shock and the shear layer.

Results comparison in terms of pressure and heat flux distribution is presented in **Figure 5**, where measures available for the heat flux are also provided.

As shown, the computed heat transfer is within the experimental uncertainty upstream of the separation point at x_{tr} = 58% L_1 (i.e., x = 27 mm); while rather good agreement with experiment is observed over the second wall of the double wedge, where the heat flux and pressure overshoots take place due to the shear layer impingement.

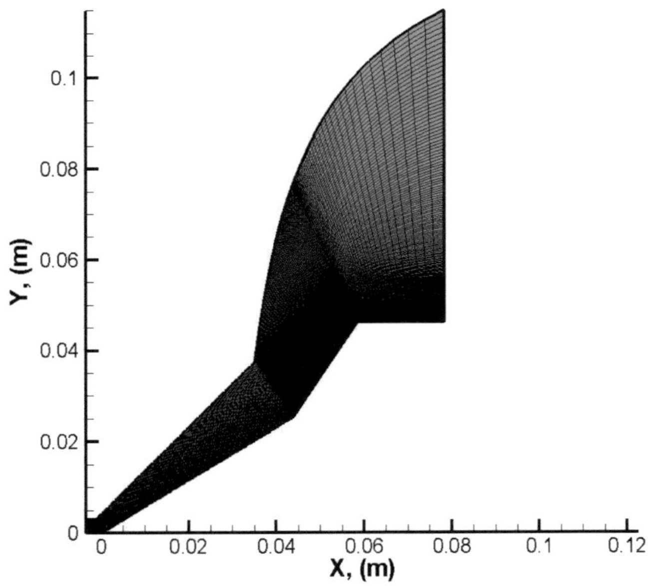

Figure 3. An example of the computational grid.

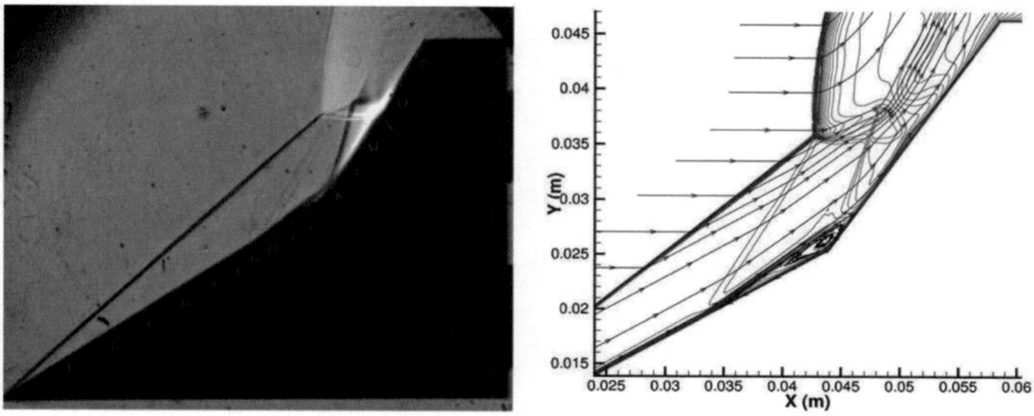

Figure 4. Comparison between Schlieren image and Mach isolines with streamtraces.

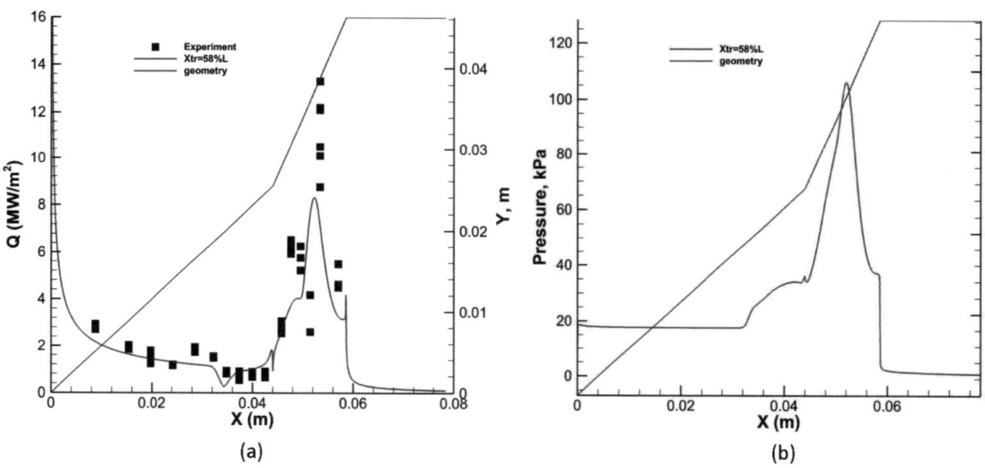

Figure 5. Heat flux and pressure profiles. Comparisons with experimental data.

In particular, **Figure 5** points out that CFD predicts well the recirculation bubble and the peak heat transfer location, but the numerical value is about 43% of that measured during the experiment [10].

Regarding pressure distribution, it is noticeable the pressure increase behind the separation shock on the first ramp. Then, a pressure overshoot, located just downstream of the reattachment point, is predicted on the second ramp. This is typical for the Edney type IV interaction. After the peak, the pressure suddenly drops toward the asymptotic pressure due to the strong expansion at the end of ramp [10].

3. Launcher aerodynamic appraisal

The launcher vehicle features a hummer head cylinder, as main body, with two boosters, see **Figure 6**. Non-dimensional aeroshape sizes are also reported in figure, being *L* the launcher

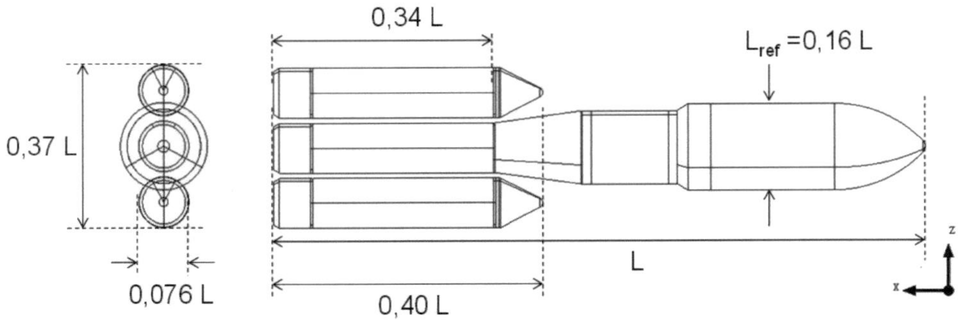

Figure 6. The launcher configuration.

height. As shown, the aeroshape under investigation also features a central core stage with a remarkable boat-tail configuration, which ends in correspondence of booster stage. The fairing diameter is 16% launcher height, while that of booster is equal to 0.076 L. The booster length is 40% of whole launcher's height [11].

Aerodynamic data for launcher are provided in the Body Reference Frame (BRF), as illustrated in **Figure 7** [11]. In this figure, aerodynamic force and moment coefficients are also provided, with sign convention according to the ISO norm. 1151.

The global aerodynamic force $\vec{F}$ and moment $\vec{M}$ acting on the launcher are expressed in BRF as follows:

$$\vec{F}=\left(-F_A\hat{i}+F_Y\hat{j}-F_N\hat{k}\right)=S_{ref}q_\infty\left(-C_A\hat{i}+C_Y\hat{j}-C_N\hat{k}\right) \tag{1}$$

$$\vec{M}=\left(M_l\hat{i}+M_m\hat{j}+M_n\hat{k}\right)=S_{ref}q_\infty L_{ref}\left(C_l\hat{i}+C_m\hat{j}+C_n\hat{k}\right) \tag{2}$$

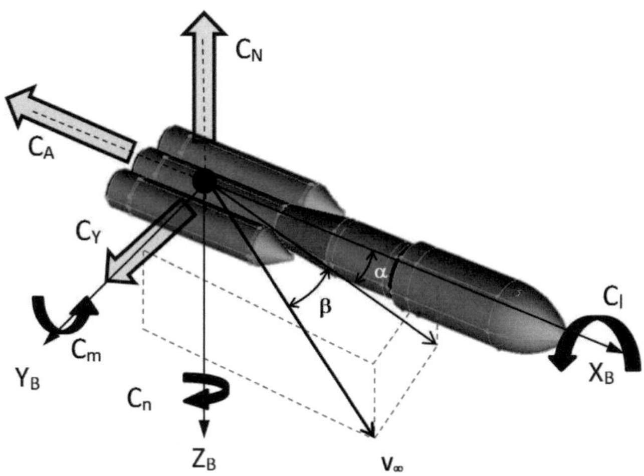

Figure 7. The body reference frame according to the ISO 1151.

where C_A is the axial force coefficient, C_Y the transverse force coefficient, C_N the normal force coefficient, $C_l = C_{Mx}$ the rolling moment coefficient, $C_m = C_{My}$ the pitching moment coefficient, $C_n = C_{Mz}$ the yawing moment coefficient, $\left(\widehat{i}, \widehat{j}, \widehat{k}\right)$ are the reference unit vectors, S_{ref} the reference surface, L_{ref} the reference length (see **Figure 6**), and q_∞ the free-stream dynamic pressure.

The definition of force and moment coefficients is:

$$C_i = \frac{F_i}{S_{ref} q_\infty} \quad i = A, Y, N \tag{3}$$

$$C_i = \frac{M_i}{S_{ref} q_\infty L_{ref}} \quad i = l, m, n \tag{4}$$

where ρ_∞ is the atmospheric density and V_∞ the speed relative to air, and the reference quantities (see **Figure 6**) are:

$$L_{ref} = 0.16L \tag{5}$$

$$S_{ref} = \frac{\pi L_{ref}^2}{4} \tag{6}$$

The present preliminary assessment, however, focuses on the longitudinal aerodynamic only, i.e., C_A, C_N and C_m are addressed. Aerodynamic coefficients are important at system level for the assessment of launcher general loading determinations, performances and, as well as, control. For instance, performances studies use the axial force coefficient C_A since this aerodynamic force opposes to the vehicle movement. Further, the launcher control needs the evaluation of the aerodynamic moment at the CoG since the control software changes the rocket's thrust direction in order to null global incidence of the vehicle, except during maneuvers. Anyway, considering that propellants are constantly consumed along the ascent flight, the CoG location is continuously changing too. Therefore, it is preferred to provide aerodynamic moments at a conventional location, namely moment reference center (MRC), see **Figure 7**. The relationship for the pitching moment coefficient evaluation, passing from MRC to CoG, reads:

$$(C_m)_{CoG} = (C_m)_{MRC} + C_N \frac{\Delta x}{L_{ref}} - C_A \frac{\Delta z}{L_{ref}} \tag{7}$$

where $\Delta x = x_{CoG} - x_{MRC}$ and $\Delta z = z_{CoG} - z_{MRC}$ are evaluated in the Layout Reference Frame (LRF), as shown in **Figure 8**.

The flow regime investigated for launcher aerodynamic appraisal during ascent encompasses subsonic, transonic-supersonic and hypersonic regimes.

In the present research effort, the range $0.5 \leq M_\infty \leq 5$ is investigated. Indeed, launcher aerodynamics has been addressed considering four Mach numbers, namely 0.5, 1.1, 2.5, and 5, at three angle of attacks, i.e., $\alpha = 0$, 5, and 7°, as summarized by the CFD test matrix in **Table 2**. Therefore, Eulerian and Navier-Stokes 3D CFD computations have been carried out on several unstructured hybrid meshes and in motor-off (i.e., without the effect of rocket plume) conditions.

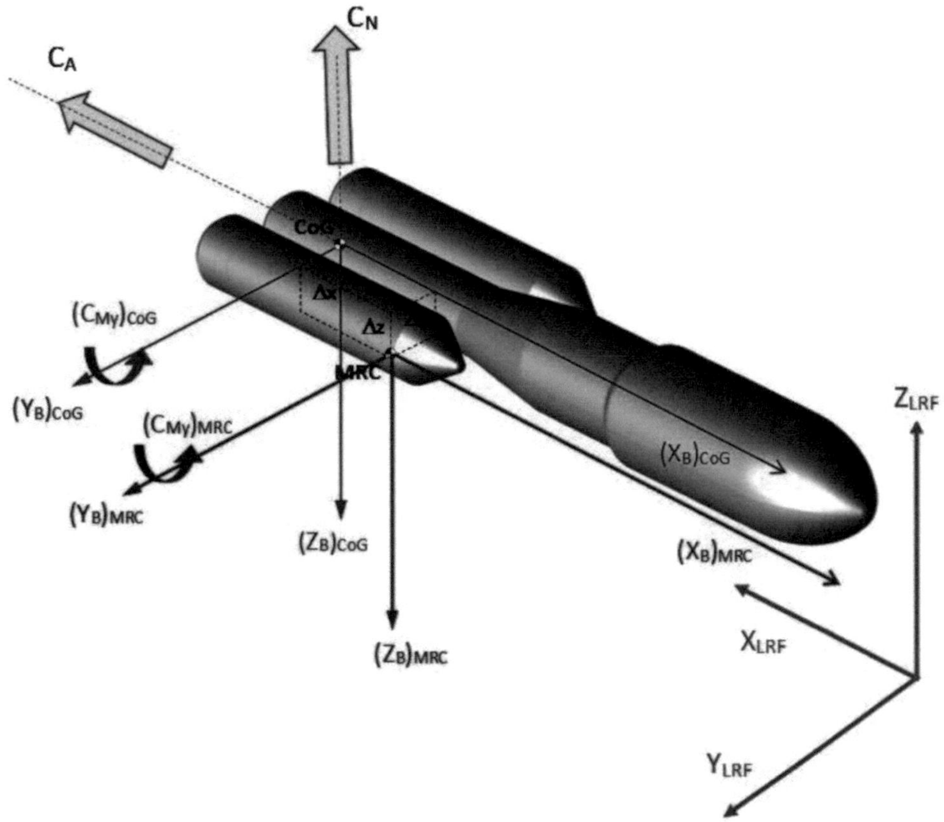

Figure 8. The layout reference frame.

α (°)	Mach			
	0.5	1.1	2.5	5
0	E	E	E	E
5	E	E	E, NS	E, NS
7	E	E	E	E

E: Eulerian CFD; NS: Navier-Stokes CFD.

Table 2. The CFD test matrix.

Engineering-based aerodynamic analyses were also performed by using a 3D Panel Method (PM) code, namely Surface Impact Method (SIM), developed by CIRA [12]. This tool is able to accomplish the supersonic and hypersonic aerodynamic and aerothermodynamic analyses of complex vehicles configuration by using simplified approaches as local surface inclination methods and approximate boundary-layer methods, respectively. Surface impact methods

http://dx.doi.org/10.5772/intechopen.70757

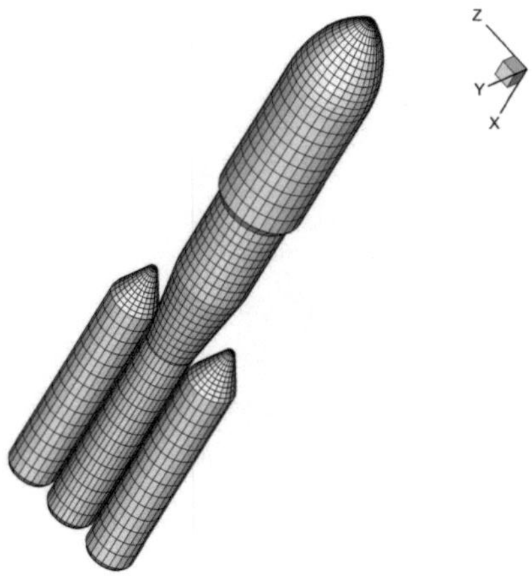

Figure 9. Panel code mesh for sup-hypersonic aerodynamics.

typical of Hypersonics, are Newtonian, Modified Newtonian, Tangent cone and Tangent Wedge theories [4].

A typical mesh surface that has been used for the engineering-level computations is shown in **Figure 9**. Some engineering-based aerodynamic results for axial and normal force coefficient are provided in both **Figures 17** and **19**, respectively.

On the other hand, the mesh domains for subsonic and sup-hypersonic speed flow simulations are shown in **Figures 10** and **11**, respectively [13].

As on can see, a square brick wide 20 body length upstream, downstream, upward and downward the launcher is considered to assure farfield unperturbed free-stream flow conditions at subsonic speed. Indeed, in this flow regime (i.e., elliptic flow), disturbances due to the

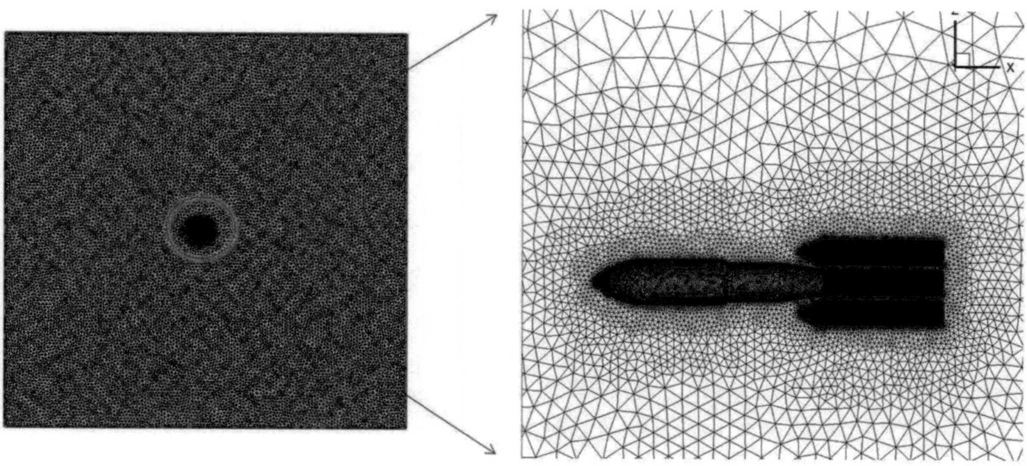

Figure 10. Overview of the hybrid mesh domain for subsonic speed.

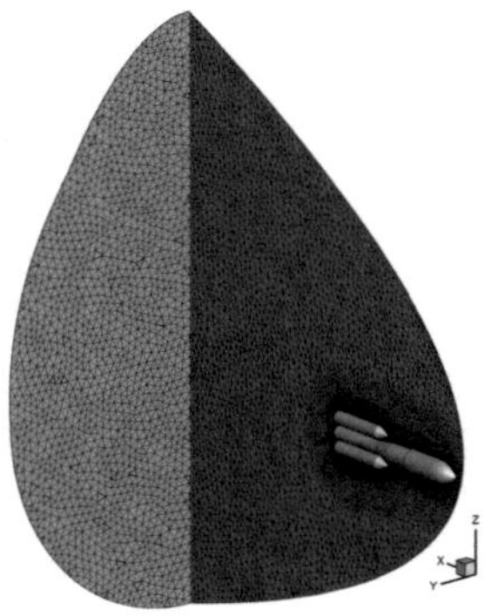

Figure 11. Overview of a hybrid mesh domain for sup-hypersonic CFD simulations.

body influence flow everywhere since they are propagated upstream via molecular collisions at approximately the speed of sound. Therefore, the computational domain must be wide enough to avoid interferences between flowfield and farfield boundary conditions.

At supersonic speed, however, a shock wave appears at launcher leading edge (i.e., hyperbolic flowfield) because of, when flow moves faster than the speed of sound, disturbances cannot work their way upstream but coalesce forming a standing wave, namely bow shock. As a result, the computational domain is quite narrow, as shown in **Figure 11**.

CFD results of the preliminary assessment of launcher aerodynamics are summarized from **Figures 12–21**. For instance, **Figure 12** shows the pressure distribution expected on the surface of the launcher flying at $M_\infty = 0.5$ and $\alpha = 5°$. Flow compression that takes place for this flight conditions at the stagnation regions of launcher fairings and of boosters' conical forebody is clearly shown. A recompression zone at the beginning of the cylindrical trunk, just after the fairings, and on that close to the boosters' forebody can be noted as well.

Results for numerical investigations at higher Mach numbers are provided in **Figure 13**. Here, an overview of pressure coefficient (Cp) distribution on launcher symmetry plane and surface is provided for $M_\infty = 2.5$ and $\alpha = 5°$.

Flow streamtraces on the symmetry plane are reported as well [13]. This CFD computation is carried out with SST k-ω turbulence flow model and for cold wall boundary condition (i.e., $T_w = 300$ K).

Results in **Figure 13** highlight a complex flowfield past the launcher due to the flow separation bubble at fairing boat-tail and the effect of fuselage/booster SSI and SWIBLI. For instance, after compression at conical flare of main fairings the flow undergoes to expansions that align it along with the constant cross section part of hammerhead. Hence, at the end of fairings another strong expansion takes place to accommodate the flow to the variation in launcher cross section (i.e., narrow cross section due to fairing boat-tail). Then, a shock wave arises at the beginning of the cylindrical trunk, just after the fairings, to redirect the flow along with the launcher wall.

Flow complexity increases further in the region close to the boosters leading edges, as also shown in **Figure 14**. This figure provides an overview of pressure coefficient distribution on launcher

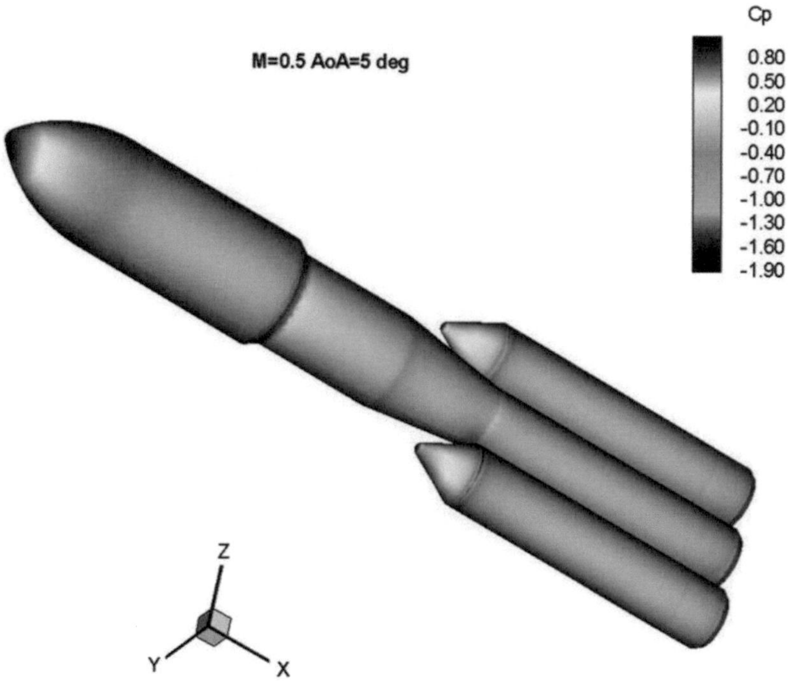

Figure 12. Pressure coefficient at $M_\infty = 0.5$ and $\alpha = 5°$.

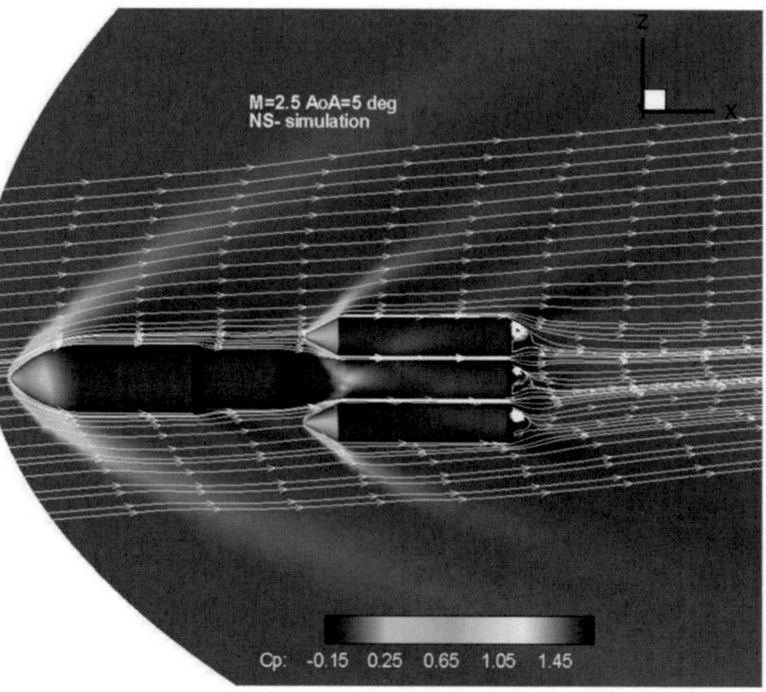

Figure 13. Overview of Cp distribution on symmetry plane and launcher at $M_\infty = 2.5$ and $\alpha = 5°$.

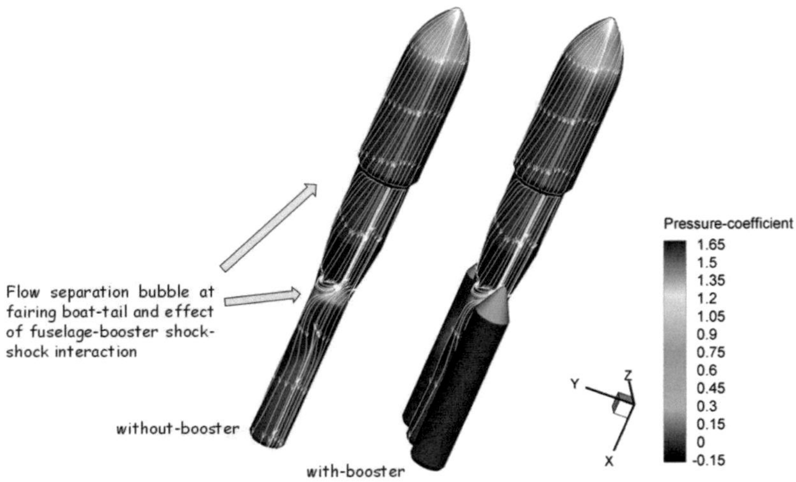

Figure 14. Contours of Cp with skin friction lines on launcher aeroshape at $M_\infty = 2.5$ and $\alpha = 5°$.

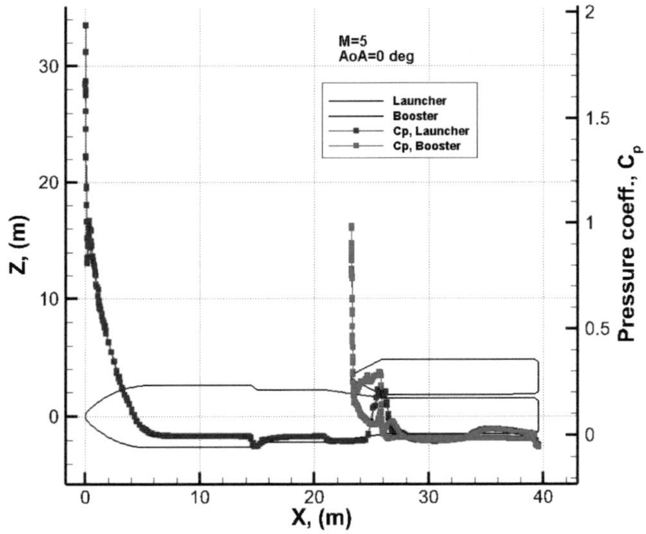

Figure 15. Profiles of Cp on launcher and booster centerlines at $M_\infty = 5$ and $\alpha = 0°$.

aeroshape with skin friction lines. As one can see, in the region close to the boosters' leading edges, complex SSI and SWIBLI phenomena take place. They result in higher thermo-mechanical loads (i.e., local pressure and thermal overshoots) on the launcher wall that must be carefully addressed in the vehicle design [13].

The effect of SSI between launcher and booster at $M_\infty = 5$ and $\alpha = 0°$ flight conditions is clearly highlighted by the pressure overshoots shown at about $x = 26$ m in **Figure 15**. As one can see, also at those flight conditions complex flowfield interaction phenomena are expected.

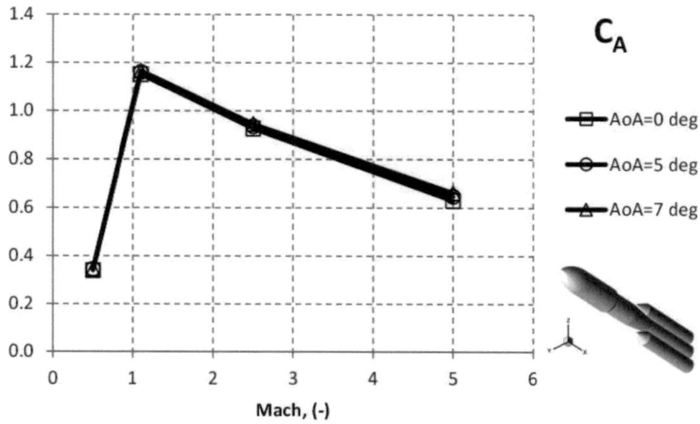

Figure 16. Axial force coefficient versus Mach at different AoA, namely $\alpha = 0$, 5, and 7°.

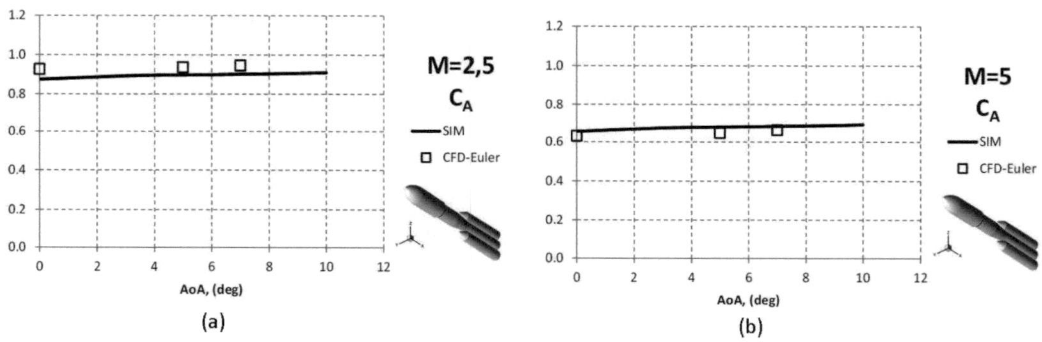

Figure 17. C_A versus AoA at $M_\infty = 2.5$ and 5. Comparison between PM and CFD results.

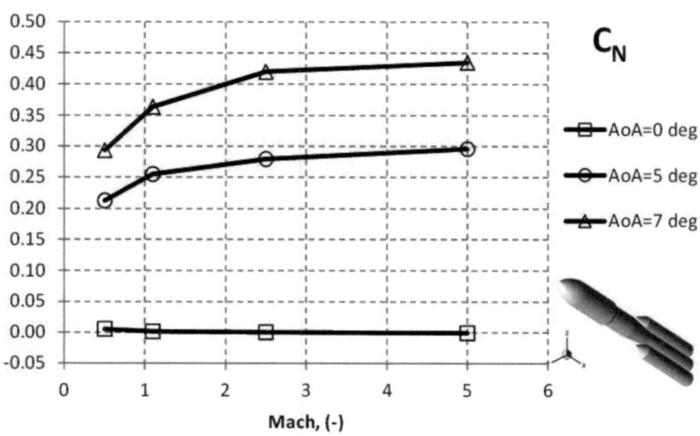

Figure 18. Normal force coefficient versus Mach at different AoA, namely $\alpha = 0$, 5, and 7°.

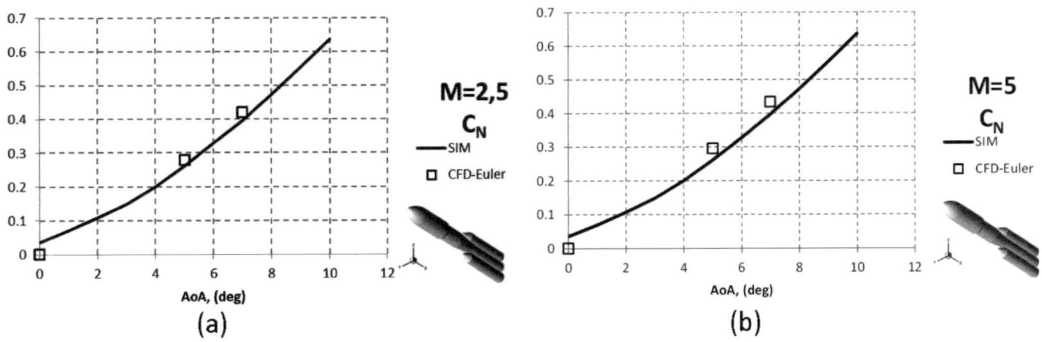

Figure 19. C_N versus AoA at $M_\infty = 2.5$ and 5. Comparison between PM and CFD results.

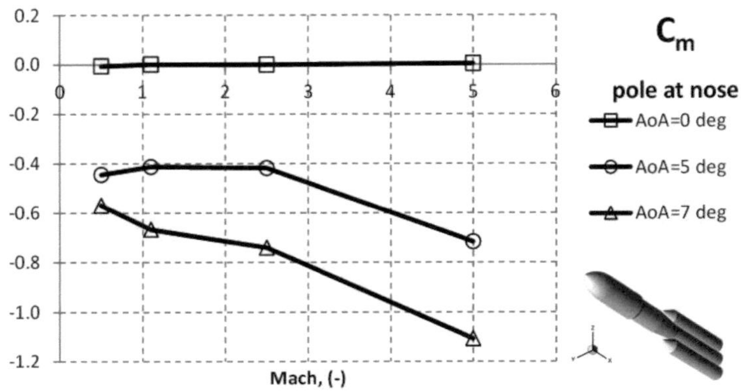

Figure 20. Pitching moment coefficient versus Mach at different AoA, namely $\alpha = 0$, 5, and 7°.

As far as aerodynamic coefficients are concerned, results for launcher axial force, normal force and pitching moment coefficients are summarized from **Figures 16–21**. For instance, **Figure 16** shows the axial force coefficient versus Mach number at different AoA, namely $\alpha = 0$, 5, and 7°. As one can see, C_A does not significantly change passing from 0 to 7° AoA at each considered Mach number.

On the contrary, the effect of flow compressibility is remarkable, as expected. Indeed, the strong increase to which undergoes the axial aerodynamic force, when M_∞ becomes transonic, is due to the wave drag contribution, as expected. Nevertheless, this contribution tends to be less strong as Mach number goes toward hypersonic speed conditions considering that the shock becomes weak due to the streamlined vehicle aeroshape (i.e., high inclined shock to assure a narrow shock layer).

The variation of C_A versus the angle of attack, α, at $M_\infty = 2.5$ and 5 is provided in **Figure 17**, where a comparison between SIM and CFD results is also available. As shown, engineering and numerical results compare rather well, thus confirming the reliability of the panel methods outcomes [14].

Regarding normal force coefficient results, **Figure 18** highlights that, for each Mach number, C_N features a quite linear slope as α increases up to 7° AoA. In addition, in this case,

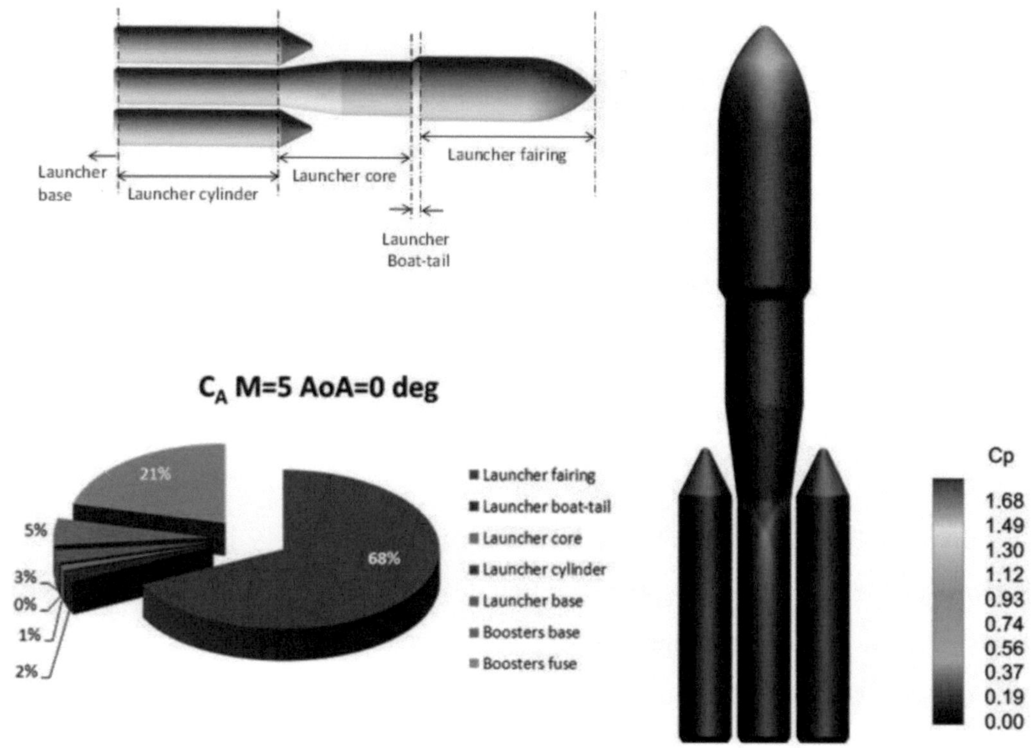

Figure 21. C_A breakdown at $M_\infty = 5$ and $\alpha = 0°$.

compressibility effect influences launcher normal force by means of different curve slopes for each Mach number. Results comparison between SIM and CFD is provided in **Figure 19** at $M_\infty = 2.5$ and 5.

As one can see, the reliability of the panel methods outcomes is still confirmed.

The vehicle pitching moment coefficient features a behavior quite close to that described for the C_N, but with a strong pitch down detected moving toward $M_\infty = 5$.

Note that both C_N and C_m at $\alpha = 0°$ are null due to the symmetric launcher aeroshape.

Finally, the axial coefficient breakdown at $M_\infty = 5$ and $\alpha = 0°$ is shown in **Figure 21**. Here the lumped contributions of launcher fairings, boat-tail, core, cylinder, and base, as well as of booster fuse and base are recognized. As one can see, launcher fairings contribute to about 68% of total drag coefficient; while this percentage for booster fuselage and base is close to 21 and 5%, respectively.

4. Conclusion

In this research effort launcher aerodynamic design activities at phase-A level are described.

The goal is to address the preliminary aerodynamic database of a typical launch vehicle configuration as input for performances evaluations as well as launcher control, sizing, and staging dynamics. To this end, both reliable engineering-based and steady-state computational fluid dynamics, with both Euler and Navier-Stokes approximations, are carried out at several Mach numbers, vehicle attitude conditions, and in motor-off conditions. In particular, launcher aerodynamic performance is provided in terms of axial, normal and pitching moment coefficients. Numerical results point out that the axial force coefficient does not significantly change passing from 0 to 7° angle of attack at each considered Mach number; while the effect of flow compressibility is remarkable. Regarding normal force coefficient, results highlight that, for each Mach number, it features a quite linear slope as the angle of attack increases up to 7°. Finally, the behavior of the vehicle pitching moment coefficient is quite close to that described for the normal force coefficient, but a strong pitch down is detected when launcher speed becomes hypersonic.

Author details

Giuseppe Pezzella[1]* and Antonio Viviani[2]

*Address all correspondence to: g.pezzella@cira.it

1 Italian Aerospace Research Centre (CIRA), Capua, Italy

2 University of Campania "L. Vanvitelli", Aversa, Italy

References

[1] Pallegoix JF. Launcher Aerodynamics. NATO: STO-EN-AVT-206; 2014

[2] Viviani A, Pezzella G. Next Generation Launchers Aerodynamics, T.C. 37/661 (2). Fort P. O., Trivandrum-695 023, Kerala, India: Research Signpost; 2012. ISBN: 978-81-308-0512-2

[3] Catalano P, Marini M, Nicolì A, Pizzicaroli A. CFD contribution to the aerodynamic data set of the Vega launcher. Journal of Spacecraft and Rockets. 2007;**44**(1):42-51 http://dx.doi.org/10.2514/1.23534

[4] Bertin JJ. Hypersonic Aerothermodynamics. AIAA (American Institute of Aeronautics and Astronautics). Washington; 1994

[5] Anderson JD. Hypersonic and High Temperature Gas Dynamics. New York: McGraw-Hill Book Company; 1989

[6] Rogers SE, Dalle DJ, Chan WM. CFD simulations of the space launch system ascent aerodynamics and booster separation. In: 53rd AIAA Aerospace Sciences Meeting, AIAA SciTech Forum, (AIAA 2015-0778) https://doi.org/10.2514/6.2015-077

[7] Viviani A, Pezzella G. Computational flowfield analysis of a next generation launcher. In: 6th European Conference for Aeronautics and Space Sciences (Eucass); 29 June-3 July 2015; Krakow, Poland

[8] Pezzella G, Marini M, Roncioni P, Kauffmann J, Tomatis C. Preliminary design of vertical takeoff hopper concept of future launchers preparatory program. Journal of Spacecraft and Rockets. 2009;**46**(4):788-799. DOI: 10.2514/1.39193 ISSN: 0022-4650

[9] Swantek AB, Austin JM. Heat transfer on a double wedge geometry in hypervelocity air and nitrogen flows. In: 50th AIAA Aerospace Sciences Meeting including the New Horizons Forum and Aerospace Exposition; 09–12 Jan 2012; Nashville, Tennessee. AIAA 2012-0284

[10] Knight D, Chazot O, Austin J, Ali Badr M, Candler G, Celik B, de Rosa D, Donelli R, Komives J, Lani A, Levin D, Nompelis I, Panesi M, Pezzella G, Reimann B, Tumuklu O, Yuceil K. Assessment of predictive capabilities for aerodynamic heating in hypersonic flow. Progress in Aerospace Sciences. 2017;**90**:39-53 http://dx.doi.org/10.1016/j.paerosci.2017.02.001

[11] Viviani A, Pezzella G. Numerical analysis of the flowfield past a next generation launcher. In: 20th AIAA Hypersonics. Strathclyde University Technology & Innovation Center; 6-9 July 2015; Glasgow, Scotland. AIAA-2015-3535. DOI: 10.2514/6.2015-3644

[12] Pezzella G. Preliminary aerodynamic and aerothermodynamic assessment of the VTO Hopper booster. ISRN Mechanical Engineering. 2011;**2011**:15. Article ID 215785. DOI: 10.5402/2011/215785

[13] Viviani A, Pezzella G, D'Amato E. Aerodynamic analysis with separation dynamics of a launcher at staging conditions. In: 30th Congress of International Council of the Aeronautical Sciences. ICAS 2016. DDC Daejeon Korea; September 25-30, 2016

[14] Pezzella G. Aerodynamic and aerothermodynamic design of future launchers preparatory program concepts. Aerospace Science and Technology. December 2012;**23**(1):233-249 http://dx.doi.org/10.1016/j.ast.2011.07.011

Section 2

Miscellaneous Topics

Chapter 5

Modeling and Analysis of Fluid-Thermal-Structure Coupling Problems for Hypersonic Vehicles

Fang Chen, Shengtao Zhang and Hong Liu

Additional information is available at the end of the chapter

http://dx.doi.org/10.5772/intechopen.70658

Abstract

To efficiently model fluid-thermal-structural problems for thermal protection design of hypersonic vehicles, a framework of Hypersonic Computational Coupling Dynamics (HyCCD) software integrates an independently developed program solving hypersonic aerthermodynamic simulation with a finite element analysis professional software. With the mathematical and physical description of multi-physics coupling mechanism, the corresponding efficient coupling strategies were proposed. Some representative coupling problems encountered in hypersonic vehicle were systematically analyzed to study the intrinsic fluid-thermal-structural coupling characteristics and mechanisms. The results can theoretically and technically support the studies on comprehensive performance assessment and optimization of thermal protection system and static or dynamic aero-thermoelastic problem of hypersonic vehicles.

Keywords: hypersonic vehicle, multi-physics coupling, aerothermodynamics, aerothermoelasticity, thermal protection system

1. Introduction

In recent years, with the ramjet engine matures and scramjet engine has achieved remarkable progress, a new-generation hypersonic vehicle powered by air-breathing engine has attracted widespread concerns in the international community and has become the focus of the future development in the field of aerospace. However, the development of this new generation air-breathing hypersonic vehicle faces many new key issues and requires great progresses in aerodynamics, structure thermal protection, propulsion technology and flight control [1, 2].

The sustained long-range maneuverable flight in the near-space atmosphere within a wide Mach range makes the experience of aerodynamic environment surrounding the hypersonic vehicles extremely complicated, which is characterized as complex flow fields, high enthalpy and long duration aeroheating with medium/low heat flux. Also, the interaction between the aerodynamic force/aeroheating flux of the external flow field and the heat transfer/thermal stress/deformation and other physical field of the internal physical field in the thermal protection system (TPS) will become extremely strong. Furthermore, the massive application of lightweight flexible materials and large thin-walled structure, especially the flight control rudder and other components will lead to another problem of aerothermoelasticity [3], which should consider the influence of sustained aeroheating. Therefore, the coupling between multi-physics such as flow field, heat and structure should be taken into account for a new-generation air-breathing hypersonic vehicle with the ability of hypersonic long-range maneuverable flight in the near-space atmosphere.

This chapter is to systematically study and analyze the coupling characteristics and mechanism of multi-physics coupling problems such as fluid-thermal-structural coupling of hypersonic vehicle, to construct a reasonable multi-physics coupling model, and to propose effective coupling analysis strategy based on computational fluid dynamics (CFD), computational thermodynamics (CTD) and computational structural dynamics (CSD), so as to provide theoretical support and analysis tools for further study of non-ablative thermal protection, aerothermoelasticity and other key issues. The following sections will focus on the modeling and analysis of several representative multi-physics coupling problems encountered on the fuselage, inlet, and wing of hypersonic vehicles.

2. The description of multi-physics coupling problem

As shown in **Figure 1**, the multi-physics coupling problem mainly involves fluid and solid in which a complex physical process takes place between aerothermodynamics within the fluid and thermo-structural dynamics within the solid through a fluid-solid coupling interface. From the view of systematic engineering, the multi-physics coupling problem constitutes a multi-physics, multi-size, and multi-variable coupling system with high coupling complexity. In the coupling system, aerothermodynamic is the active motivation including the coupling of aerodynamic force and aerodynamic heat, while the thermo-structural dynamics is the passive response including the coupling of heat transfer, thermal stress and deformation.

The high complexity of multi-physics coupling problems makes it very difficult to establish a complete coupling model simultaneously considering all the coupling relations and factors. It is necessary to split the problem into different coupling levels according to physical environment characteristics and the engineering application background. For the blunt leading edge and the fuselage with large heat insulation areas with large rigidity, the structural deformation is relatively weak and can be ignored so that the multi-physics coupling problem can reduce to fluid-thermal coupling problem or fluid-solid conjugate heat transfer problem. However, the structural deformation of some structures

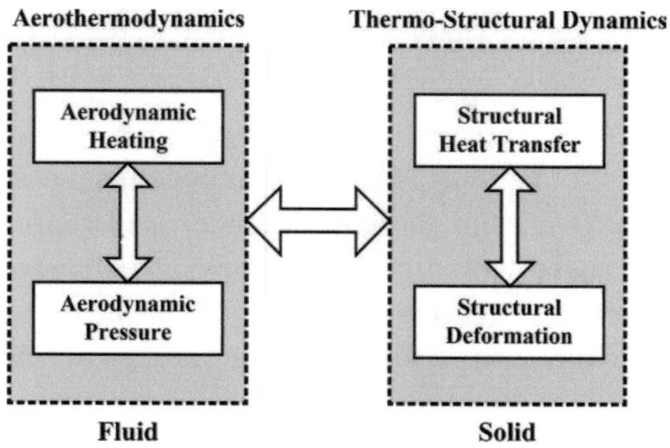

Figure 1. Coupling mechanism of multi-physics coupling problems for hypersonic vehicle.

with low rigidity is no longer negligible and induces thermal stress and thermal deformation while the fluid-thermal-structural coupling problem mainly characterized as the coupling between aerodynamic force/heat and heat transfer/thermal stress/deformation. In particular, the aerothermoelastic problem behaves more prominently for the large thin-walled flexible structures such as the wings and flight control rudders in which the fluid-thermal-structural coupling should consider the inertial effect and vibration of the structures.

The modeling of so-called multi-physics coupling problem mainly refers to constructing the mathematical-physical model to describe the coupling behavior of the multi-physical fields, namely, the partial differential equation systems (PDEs) to describe the multi-physics coupling problem and the corresponding initial/boundary conditions. And then, the analysis is to solve the partial differential equations by numerical simulation method to obtain the physical properties and behaviors. This modeling and analysis can generally be divided into two different types [4, 5], that is, the monolithic coupling approach and the partitioned coupling approach. According to the characteristics of multi-physics coupling problems, the global strategy for modeling and analysis is shown in **Figure 2**. The monolithic coupling approach is used respectively for the aerodynamic force/heat coupling within the fluid and the thermo-structural dynamic problems within the solid. In contrast, the partitioned coupling approach is applied for the fluid-thermal coupling and fluid-thermal-structural coupling problem through the fluid-solid coupling interface.

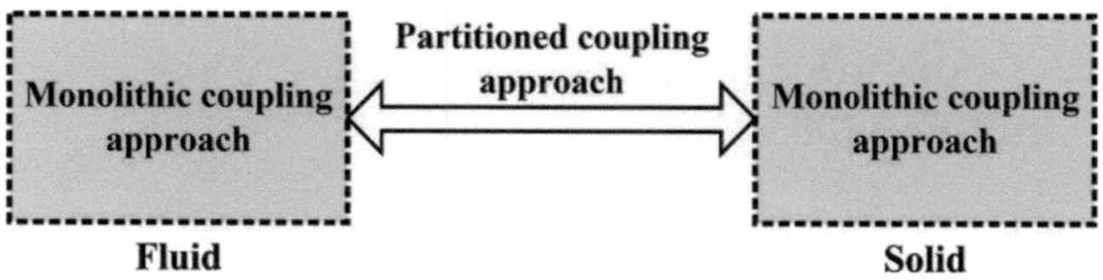

Figure 2. Global strategy for modeling and analysis approaches.

3. Modeling and analysis of flow-thermal coupling problem

The coupling mechanism of fluid-thermal coupling problem (or fluid-solid conjugate heat transfer problem) is a physical process of interaction between aeroheating within the fluid and the heat transfer within the solid through the fluid-solid coupling interface. When the vehicle flies at hypersonic speed within the atmosphere, it will face strong aeroheating due to strong shock compression and viscous friction. A part of aerodynamic heat flux radiates from the surface while the remaining transfers to the internal structure. The heat into the solid structure is closely related to the structure layout, material properties and various heat transfer boundaries of radiation and convection within the solid, thus forming specific heat distribution characteristics within the solid structure in the form of transient temperature field distribution. Meanwhile, the heat distribution characteristics within the solid structure, in turn, also restricts aeroheating further entering the internal solid structure through the variation of the wall temperature. It represents a strong two-way coupling relationship between the aeroheating environment of the external flowfield and the thermal response of internal solid structure. The aeroheating environment as thermal load is active excitation, which changes continuously along the flight trajectory so that the fluid-thermal coupling problem appears as a sustained non-transient physical coupling process.

This coupling process involves three different time scales, that is, the characteristic time of the dynamic flight trajectory, the characteristic time of the flow response and the characteristic time of the structural thermal response. Research and development of the coupling analysis strategy should take full account of the obvious difference in the above time scales. Two concepts are introduced herein: (1) static flight trajectory, which refers to the flight state (height, speed and angle of attack) remaining constant over time; and (2) dynamic flight trajectory, which refers to the flight state (height, speed and angle of attack) changing continuously over time.

3.1. Coupling analysis strategies based on static flight trajectory

3.1.1. Loosely-coupled analysis strategy

For hypersonic fluid-thermal coupling problems, in most cases, the characteristic time scale of structural thermal response is much longer than that of the flow response. Thus, the flow response can be assumed to be frozen over time compared with the structure thermal response. In other words, when the hypersonic flow is disturbed at certain time, a steady state can be reached instantaneously without relaxation process. Based on this physical assumption, the partitioned coupling method by calculating the steady flowfield and the transient structure heat transfer respectively is a good approximation, which can greatly improve the computational efficiency of coupling analysis.

A loosely-coupled analysis strategy for hypersonic fluid-thermal coupling problem based on static flight trajectory [6] is shown in **Figure 3**. Δt_{F} is the flowfield calculation time step, Δt_{T} is the structure heat transfer calculation time step. Δt_{C} is the coupling calculation time step, which can be set as several times of the structure heat transfer calculation time step, that is $\Delta t_{\mathrm{C}} = \mathrm{n} \cdot \Delta t_{\mathrm{T}}$ $(n = 1, 2, 3, \cdots)$. The interface coupling relations are achieved by the

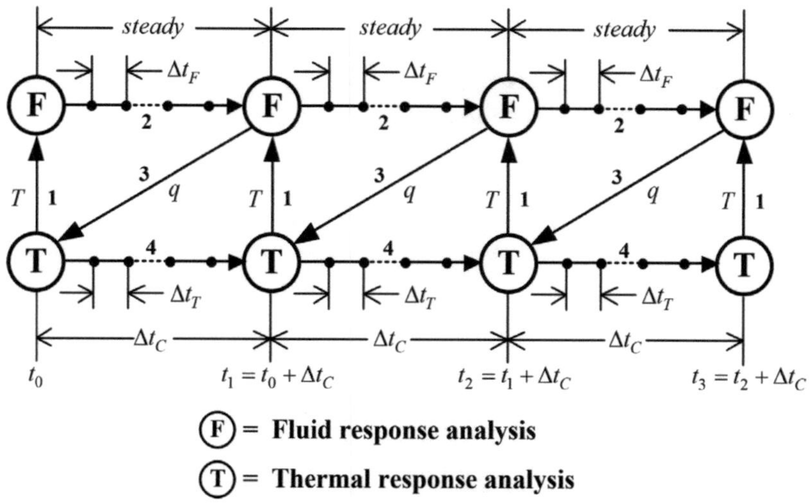

Figure 3. Loosely-coupled analysis strategy for fluid-thermal coupling problems.

Dirichlet-Neumann model [7], which has been validated for the fluid-thermal coupling problem solved by partitioned coupling method.

The loosely-coupled analysis strategy can be summarized as follows:

1. At the initial time t_0, an initial constant temperature or temperature field distribution is firstly given to the solid structure, and then the temperature distribution of the solid structure on the fluid-solid coupling interface is transferred to the fluid domain as a Dirichlet boundary condition of the flowfield calculation by the interface information transfer method.

2. By calculating the steady flowfield within the fluid domain based on the imposed boundary condition of temperature, wall heat flux distribution of the steady flowfield can finally be obtained.

3. The wall heat flux distribution of the steady flowfield is transferred to the solid domain as the Neumann boundary condition for the heat transfer calculation of the solid structure by the interface information transfer method.

4. Calculation of transient heat transfer from t_0 to $t_0 + \Delta t_C$ within the solid domain can be done based on the imposed heat flux boundary conditions to obtain the solid structure temperature field distribution at $t_0 + \Delta t_C$.

5. The solid structure wall temperature distribution at time $t_0 + \Delta t_C$ is transferred to the fluid domain as the Dirichlet boundary condition for the flowfield calculation by the interface information transfer method.

6. When the calculation in one coupling time step has been completed, the calculation will continue in the next time step until all the time steps are covered.

3.1.2. *Tightly-coupled analysis strategy*

In the loosely-coupled analysis strategy, only one information exchanges between the fluid and the solid in each coupling calculation time step, and thus the coupling calculation is relatively efficient. However, this loosely-coupled analysis strategy does not strictly satisfy time synchronization of interface coupling relations, which reduces the accuracy of coupling calculation. On the basis of the above loosely-coupled analysis strategy, the sub-iteration strategy can be introduced to each coupling calculation time step, which allows multiple information exchanges between the fluid and the solid. The time marching calculations of flowfield and structure heat transfer are repeated until the convergence, and then calculation will continue in the next coupling time step, which forms a tightly-coupled analysis strategy. Obviously, the tightly-coupled analysis strategy improves the accuracy of the coupling calculation, but it requires additional sub-iteration calculation which decreases the calculation efficiency.

Figure 4 shows the tightly-coupled analysis strategy of hypersonic fluid-thermal coupling problem based on static flight trajectory. The tightly-coupled strategy adds a step to repeat step (2) to step (5) until the convergence is satisfied.

3.2. Coupling analysis strategy based on dynamic flight trajectory

If the flight trajectory of the vehicle varies continuously over time, the influence of the flight trajectory variation should be considered in the coupling analysis. As the flight trajectory variation is the macroscopic motion of the vehicle as a rigid body and the time span of the continuous flight is large, the flight trajectory variation over time is relatively slow. Therefore, the time scale of affecting aeroheating characteristics is much larger than the coupling time scale between the flowfield and the heat transfer. A coupling analysis strategy for hypersonic fluid-thermal coupling problem based on dynamic flight trajectory is then shown in **Figure 5**.

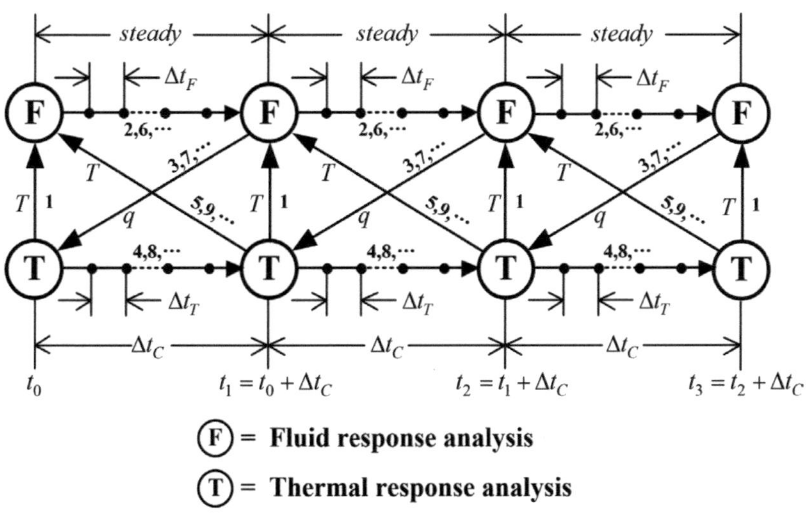

Figure 4. Tightly-coupled analysis strategy for fluid-thermal coupling problems.

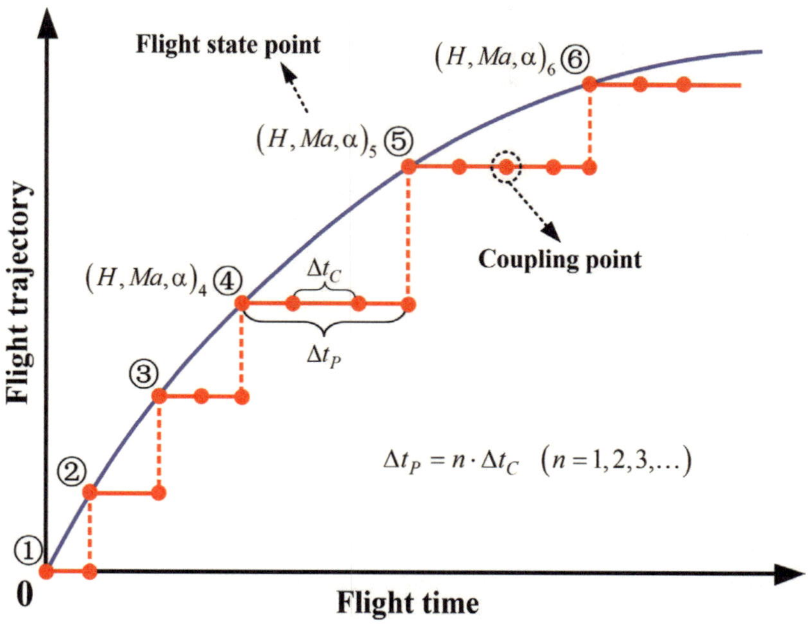

Figure 5. Coupling analysis strategy based on dynamic trajectory.

It is to discretize the fluid-thermal coupling problem of the sustained dynamic flight trajectory into a set of fluid-thermal coupling problems of quasi-static flight trajectories in chronological order. The details are listed below:

1. The flight trajectory of the vehicle is regarded as a continuous flight state function in the time domain. The appropriate time step Δt_P is selected to discretize the flight trajectory into a series of discrete flight state point $(H, Ma, \alpha)_i$ and the duration of each discrete flight state point is in which the internal flight conditions remain constant, set as the average of flight conditions at the starting point and at the ending point of the discrete flight state.
2. During Δt_P of each discrete flight state, the problem is regarded as the fluid-thermal coupling problem based on the static flight trajectory. The appropriate coupling calculation time step $\Delta t_C = \Delta t_P / n$ $(n = 1, 2, 3, \cdots)$ is selected.
3. After the coupling calculation of one discrete flight state point, the coupling calculation of the next discrete flight state point in time order will be done until the end of the entire flight.

3.3. Fluid-thermal coupling analysis of hypersonic vehicle

The geometry of hypersonic vehicle model is shown in **Figure 6**. The head part is made of C/C composites, and the rest of fuselage is made of TB6 titanium alloy. The integrated analysis program platform Hypersonic Computational Coupling Dynamics (HyCCD) for

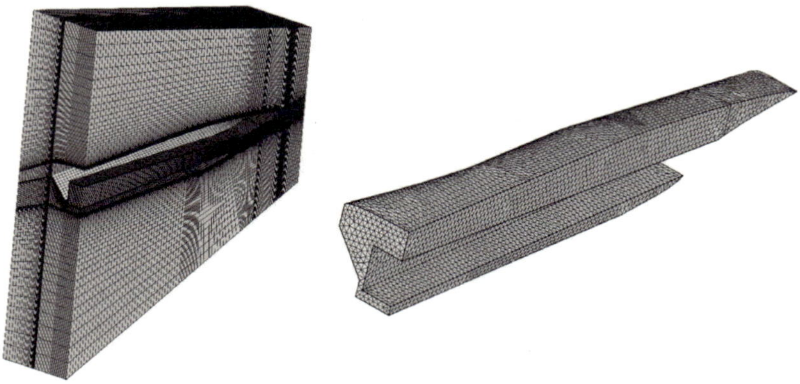

Figure 6. The geometric shape and CFD/FEM computational grid.

hypersonic fluid-thermal coupling problem is realized by integrating hypersonic aerothermodynamic numerical simulation program Hypersonic Computational Fluid Dynamics (HyCFD) and the finite element heat transfer analysis software (ANSYS Mechanical APDL). The external CFD computational grid is multi-block structured grid, the internal FEM grid is unstructured grid.

The actual flight of hypersonic vehicles usually includes climbing, cruise and descending. A simple flight trajectory is assumed here, as is shown in **Figure 7**. The flight time along the trajectory is 210 s, 0–50 s for climbing, 50–150 s for cruise and 150–210 s for descending. The discrete time step of the selected flight trajectory is $\Delta t_P = 5$ s, and the flight trajectory is divided into 42 discrete flight states. The loosely-coupled analysis strategy is used for the fluid-thermal coupling analysis, and the fluid-solid coupling time step is $\Delta t_C = 5$ s. The chemical non-equilibrium gas model is used for flowfield calculation. Non-catalytic wall is selected as the boundary condition. The initial temperature is 300 K, the inner wall temperature keeps at 300 K during the flight, and the outer wall emissivity is $\varepsilon = 0.8$.

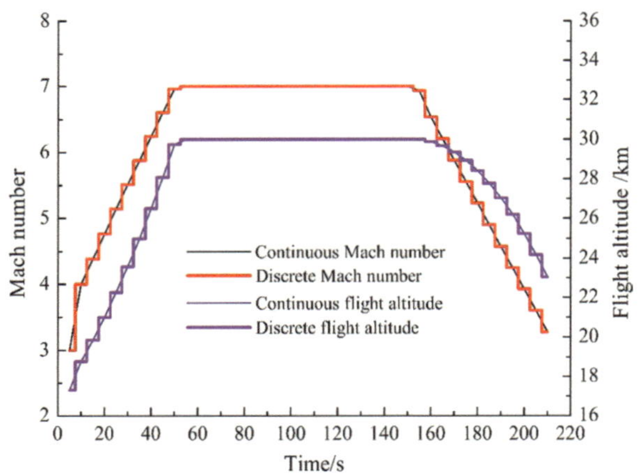

Figure 7. The dynamic flight trajectory and its discretization.

Figure 8 shows the wall temperature at the leading edge stagnation point of the symmetric plane along the dynamic trajectory. The wall temperature evolution without considering the wall radiation effect is also given in the figure for comparison. It can be seen that: (1) from the starting point $t = 0$ s to the cruise state $t = 50$ s, the wall temperature of the structure gradually increases while the aeroheating capacity of the external flowfield rises gradually due to the increasing flight speed, and thus more and more aerodynamic heat is transferred into the internal solid structure through the interface. The solid structure keeps storing heat; (2) during the cruise $t = 50–150$ s, although the flight conditions no longer change and the aeroheating capacity of the external flowfield remains constant, aerodynamic heat is still continuously transferred into the internal structure due to the fluid-solid heat transfer coupling effect. The solid structure still keeps storing heat. Meanwhile, as the wall temperature continues to rise, less and less aerodynamic heat will be transferred into the internal structure. If the cruise flight time is long enough, the fluid-solid heat transfer coupling will eventually reach equilibrium; and (3) in descending $t = 150–210$ s, the flight speed begins to decrease, the aeroheating capacity begins to weaken while the wall temperature has been extremely high, reaching the temperature peak along the entire trajectory. The solid structure begins to release heat, and thus the wall temperature decreases gradually.

Figure 9 show the temperature distribution of I-I cross-section (see **Figure 8**) within the solid structure, intuitively presenting how the temperature distribution within the structure changes with the dynamic trajectory. It should also be noted that the heat distribution and transmission within the structure will change with the trajectory of the vehicle, thus exhibiting dynamic multi-field coupling spatial-temporal characteristics. It is an effective

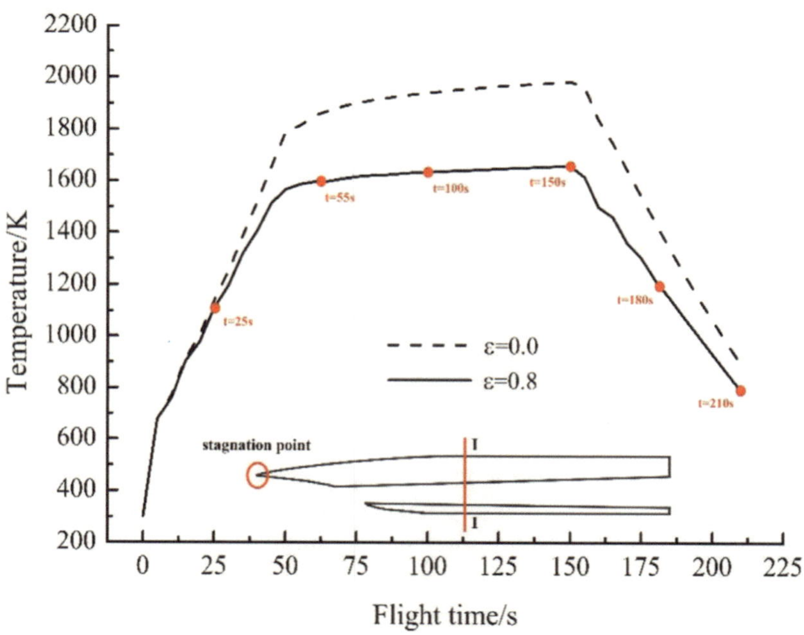

Figure 8. The wall temperature at stagnation point of the symmetric plane leading edge.

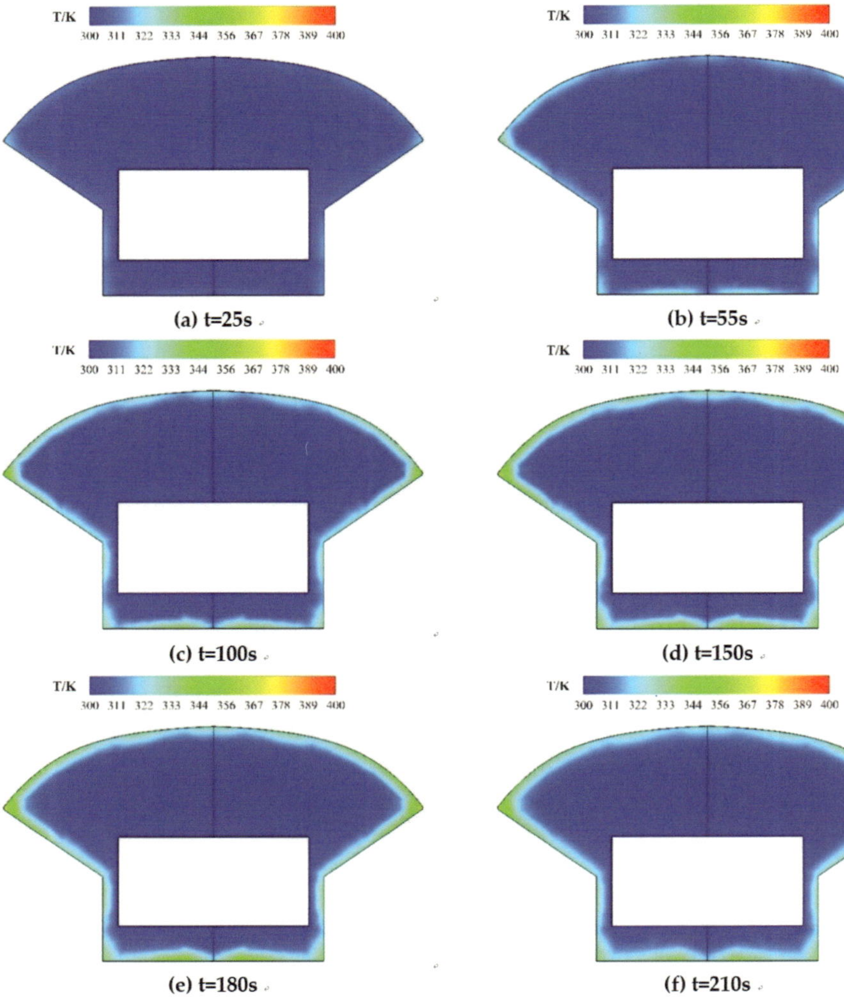

Figure 9. The temperature distribution of the I-I section plane within the solid structure. (a) $t = 25$ s, (b) $t = 55$ s, (c) $t = 100$ s, (d) $t = 150$ s, (e) $t = 180$ s, and (f) $t = 210$ s.

to optimize the integrated design of thermal protection and to study new concepts and methods of thermal protection by accurately predicting and analyzing the temporal and spatial distribution characteristics and the transmission of heat within the vehicle solid structure for guidance.

The preliminary analysis and research on coupling characteristic and influencing factors of the hypersonic fluid-thermal coupling problem reveal the spatial-temporal distribution characteristics of the fluid-solid heat transfer coupling and the influence of wall radiation effect on the sustained flight conditions. There is a close coupling between the aeroheating of the flowfield and the heat transfer of the structure. The heat distribution and transfer within the structure change with the dynamic trajectory, which displays the spatial and temporal characteristics of multi-physics coupling. The integrated analysis method and the program platform

HyCCD can effectively predict and analyze the thermal response characteristics and principals of the solid rigid structures inside the hypersonic vehicle under sustained flight conditions. The accurate prediction and analysis will be an effective way to optimize thermal protection design and to study new concepts and methods of thermal protection system of hypersonic vehicles.

4. Modeling and analysis of fluid-thermal-structural coupling problem

During the sustained flight of an actual hypersonic vehicle within the atmosphere, the strong two-way coupling of aerothermal environment with thermal response within solid structures causes the heat distribution in the form of transient temperature field. It not only has dynamic effects on the properties of solid structures also causes thermal stress due to the temperature gradient. Meanwhile, the resultant thermal strain from the thermal stress influences the heat distribution through the deformation of the solid structures. The large structural deformation even affects the aerodynamic forces/heat in the external flowfield. On the other hand, the aerodynamic forces lead to the structural stress and structural within solid structures, which can also affect aerodynamic forces/heat in the external flowfield in the form of structural deformation.

The fluid-thermal-structural coupling model is shown in **Figure 10**. The volumetric coupling of the aerodynamic forces and aerodynamic heat of the flowfield within the fluid is described by unified fluid governing equations, which is solved by computational fluid dynamics (CFD) to obtain the parameters of aerodynamic forces/heat. The thermal load (wall heat flux q) and force load (wall pressure p) are imposed on the solid through the fluid-solid coupling interface. Within the solid, the thermal response is described by governing equations of heat conduction, while the stress/strain are described by governing equations of thermoelastics.

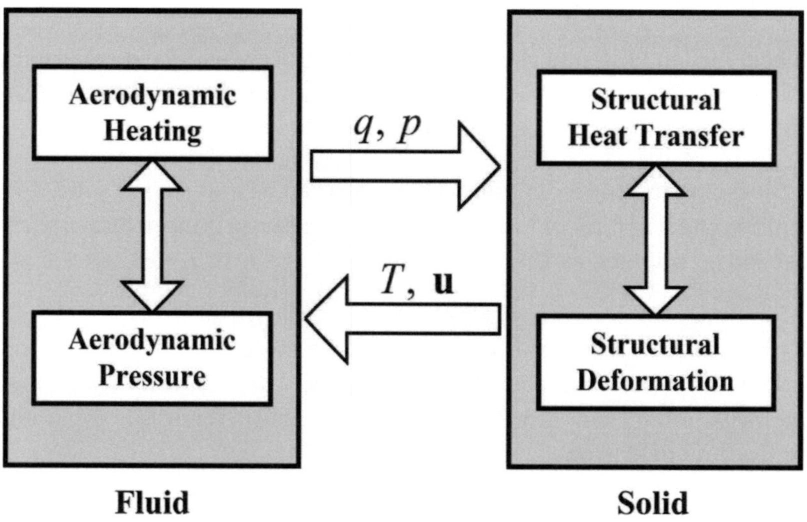

Figure 10. Fluid-thermal-structural coupling model.

Considering the effects of the temperature-deformation coupling, the parameters within the solid can be obtained by solving the governing equations of heat conduction and thermoelastics with HyCCD platform. The temperature condition (the wall temperature T) and structural deformation condition (the surface displacement $\mathbf{u}$) are provided for the fluid through the fluid-solid coupling interface.

4.1. Coupling analysis strategies based on static flight trajectory

4.1.1. Loosely-coupled analysis strategy

The loosely-coupled analysis strategy for the fluid-thermal-structural coupling problem on the basis of the static trajectory is shown in **Figure 11**. Δt_F is the time step in flow-field calculation, Δt_{TS} is the time step in thermal-structure volumetric coupling calculation of solid; Δt_C is the time step in fluid-solid coupling surficial calculation and can be set as several times the time step in thermal-structure volumetric coupling calculation of solid, that is $\Delta t_C = n \cdot \Delta t_{TS}$ ($n = 1, 2, 3, \cdots$).

The loosely-coupled analysis strategy can be summarized as:

1. At the initial time t_0, an initial constant temperature or temperature field distribution as well as the initial load and displacement constraints is given to the solid structure first. Then the wall temperature and displacement of the solid structure are transferred to the fluid domain by the information transfer method of the interface. The wall temperature is used for the boundary condition in the flowfield calculation, while the displacement is used to update the flow-field grid;

2. By calculating the steady flowfield in the fluid domain based on the imposed boundary condition of temperature and the updated flow-field grid, the wall heat flux and the wall pressure can finally be obtained;

3. The wall heat flux and the wall pressure of the steady flowfield are transferred to the solid domain as the heat load and force load respectively for the thermo-structural dynamic calculation of solid by the information transfer method of the interface;

4. Transient thermo-structural dynamic calculation can be done in the solid domain based on the imposed heat and force load to obtain the response parameters of the solid structural heat/force coupling as the time advances from t_0 to $t_0 + \Delta t_C$ and finally reaches $t_0 + \Delta t_C$;

5. The wall temperature and displacement of the solid structure at $t_0 + \Delta t_C$ are transferred to the fluid domain by the information transfer method of the interface. The wall temperature is used for the boundary condition for flowfield calculation, while the displacement is used to update the flow-field grid;

6. When the calculation in one coupling time step has been completed, the calculation will continue in the next time step until all the time steps are covered.

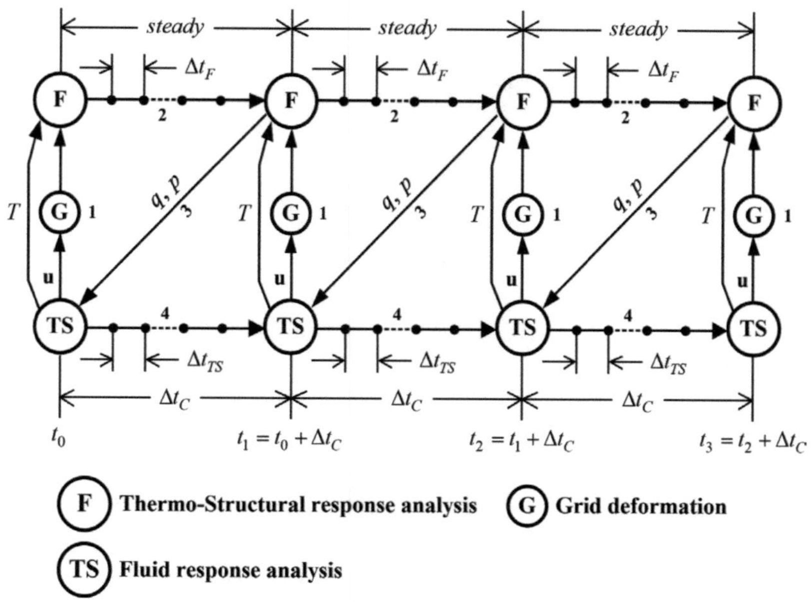

Figure 11. Loosely-coupled analysis strategy for fluid-thermal-structural coupling problem.

4.1.2. Tightly-coupled analysis strategy

By introducing sub-iteration into each computational time step of coupling in the loosely-coupled analysis strategy, the tightly-coupled analysis strategy for fluid-thermal-structural coupling problems based on the static trajectory is shown in **Figure 12**.

4.2. Coupling analysis strategies based on dynamic trajectory

The coupling analysis strategy is to discretize the sustained dynamic trajectory into a set of a finite number of quasi-static trajectories in chronological order.

4.3. Fluid-thermal-structural coupling analysis of inlet cowl leading edge

As for the fuselage-engine-integrated design, the waverider forebody is utilized for pre-compression in order to produce lift and at the same time obtain the flow rate required by the engine inlet. In this case, both the waverider forebody and the cowl leading edge are on windward side where the most severe aeroheating takes place. As shown in **Figure 13**, the forebody precompression oblique shock and the cowl leading edge shock may intersect with each other, which leads to shock interaction and thus aggravates aeroheating near the cowl leading edge, even more severe than that at the nose leading edge. Case 1 is defined as over-ideal state, in which the incident shock enters the cowl and the cowl leading edge is under the far-field freestream condition. Case 2 is defined as ideal state, in which the incident shock arrives exactly at the inlet cowl and interacts with the shock at the cowl leading edge; Case 3 is defined

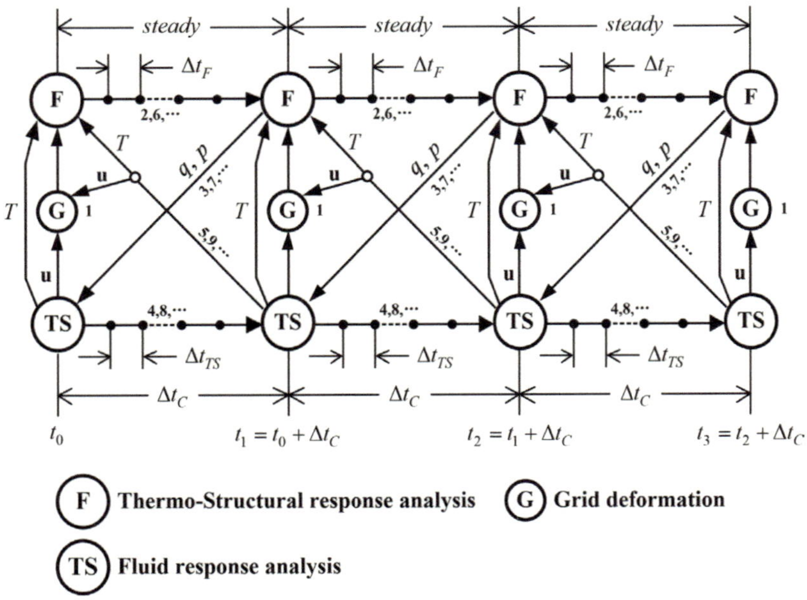

Figure 12. Tightly-coupled analysis strategy for fluid-thermal-structural coupling problem.

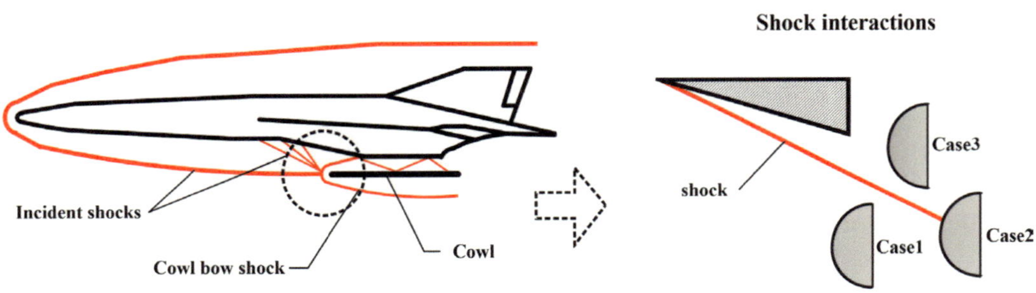

Figure 13. The shock interaction phenomena near the inlet cowl leading edge.

as under-ideal state, in which the incident shock outside has not reached the cowl and the cowl leading edge is under the downstream shock freestream condition.

The cylindrical leading edge model is used as the inlet cowl leading edge model, which utilize titanium alloy (Ti-6Al-2Sn-4Zr-2Mo) as the material. Sustained coupling calculation time of 11 seconds is selected for the fluid-thermal-structural coupling calculation and analysis of the engine cowl leading edge model; the loosely-coupled analysis strategy is employed for calculation and its fluid-solid surficial coupling calculation time step adopts the adaptive strategy. High temperature chemical non-equilibrium gas model is adopted for the calculation of the external fluid domain, non-catalytic wall is selected as the wall catalytic condition. The initial temperature for calculation of thermal-structural coupling within the solid domain is 300 K with zero initial stress and the reference temperature of thermal stress is 300 K. The aerodynamic force/thermal load of the external flowfield are taken into consideration and the wall radiation effect is also considered with the surface emissivity $\varepsilon = 1.0$ in the coupling calculation and analysis. A pressure load $p_\infty = 1197$ Pa is imposed on the inner wall, the fixed support is selected for both ends of the model.

It can be intuitively seen from **Figure 14** that under the effect of the striking of the extremely densified heat flux due to shock interaction, the heat rapidly accumulates within the structure nearby the struck point, causing a leading increase in the temperature of the point. As the heat accumulates at the point over time, the structure temperature is also increasing and at the same time the heat is gradually transferred to internal area in depth. Hence, the temperature distribution is also gradually expanding from the struck point to internal structure area in depth. Simultaneously, the amplitude of overall temperature distribution within the structure declines much if wall radiation effect is taken into account because it effectively limits the heat entering the internal structure. **Figure 15** shows that the earliest stress concentration occurs within the structure near the struck point on the wall. The stress distribution is also gradually expanding to the internal structure area in depth over time. Simultaneously it is also shown that the amplitude of overall stress distribution inside the structure declines if wall radiation effect is taken into account.

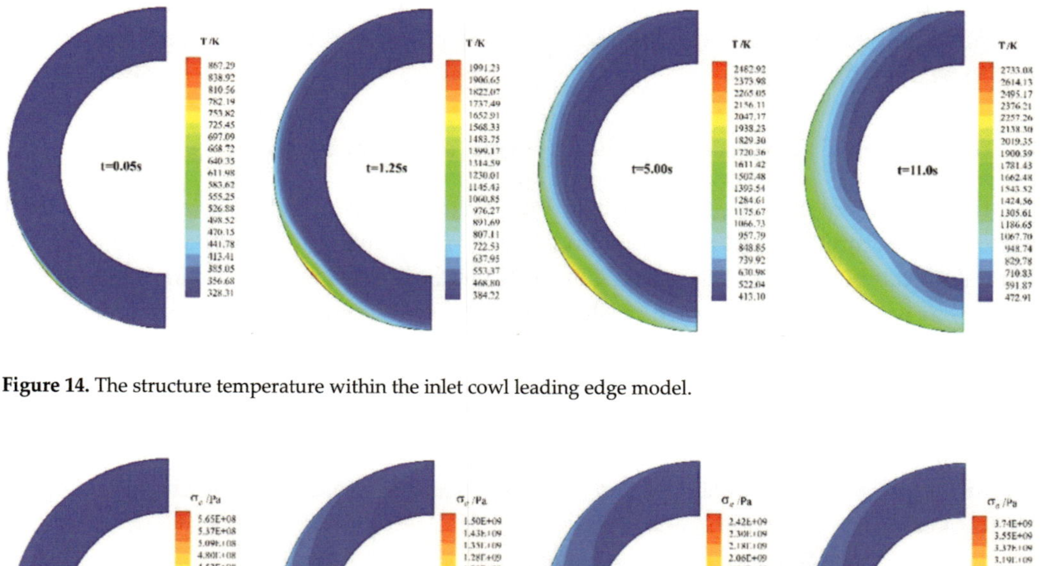

Figure 14. The structure temperature within the inlet cowl leading edge model.

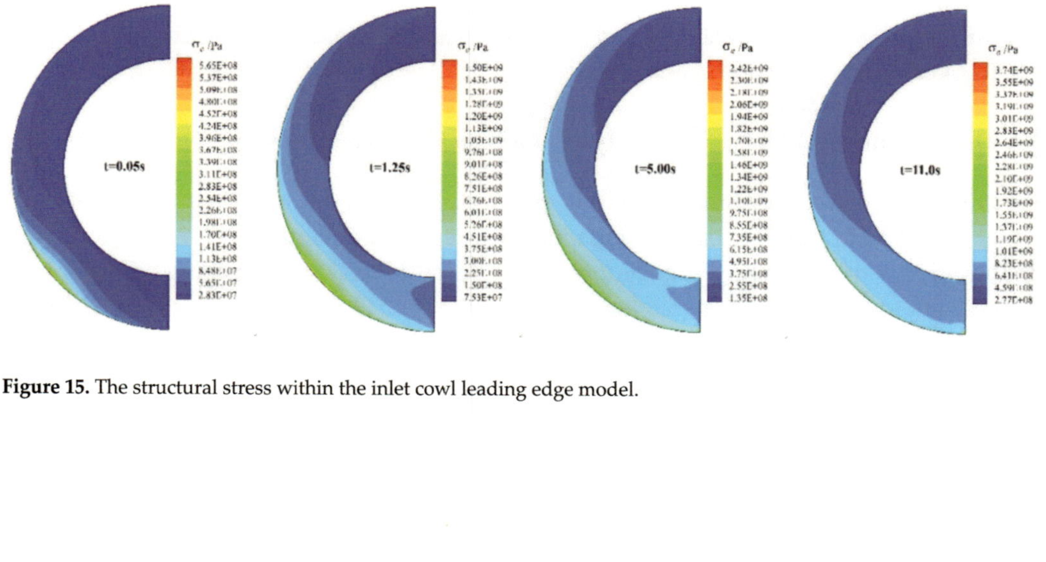

Figure 15. The structural stress within the inlet cowl leading edge model.

In general, the aerodynamic force and thermal load have a great impact on the inlet cowl leading edge, which suffices to cause thermal and dynamic damage to thermal protection structures despite the short imposing time in actual flight, presenting severe challenge for material selection and structure design of thermal protection. As for the air-breathing hypersonic vehicles, the impact of shock interaction is common in the surrounding flowfield. Therefore, thermal protection design of local leading edges (structures such as the tail and rudders) on the windward side wrapped by the nose shocks should be done carefully besides the nose and the engine cowl leading edge.

5. Research on structural thermal modals

Aeroelastic problems have been the key in vehicle design, which has gradually become a notable obstacle to better vehicle performance with the development of vehicle. Under the effect of aeroheating, the temperature rise of a structure leads to variation in physical parameters of its material. Also, non-uniform temperature field within the structure causes prominent temperature gradients, which produces subsidiary thermal deformation and thermal stress, greatly alters the structural rigidity and thus changes the natural vibration performance of the structure. The variation of natural vibration performance due to thermal load significantly affects the trim, flutter and control characteristics of the vehicle and these effects tend to be unfavorable.

5.1. Thermal modal analysis strategy based on multi-physics coupling

The thermal modal analysis strategy based on multi-physics coupling integration method is shown in **Figure 16**. It can be summarized as: (1) employing the multi-physics coupling integration method HyCCD based on CFD, CTD and CSD, transient temperature field and stress field within the solid structure along its static or dynamic trajectory are obtained by doing hypersonic fluid-thermal-structural coupling analysis according to the coupling analysis strategy described above and (2) thermal rigidity matrix can be constructed by taking parameters of thermodynamic state within the solid structure of each time point for coupling calculation. Then, the thermal modal characteristics at each time point for coupling calculation of the solid structure are obtained by solving the generalized eigenvalue problem by means of conventional mode analysis method.

5.2. Thermal modal characteristic analysis of a typical hypersonic wing

The thermal modal characteristic analysis under sustained flight of a typical three-dimensional low-aspect-ratio hypersonic wing model is presented in **Figure 17**. It is a symmetrical double edge with the leading edge blunted and small thickness of the trailing edge retained to avoid sharp edges. The material has mass density of 4539 kg/m^3 and Poisson number of 0.32 with all the other physical properties varying with temperature. The temperature at initial time is 300 K with zero initial stress and the reference temperature of thermal stress is 300 K. Fixed support is adopted at the wing root.

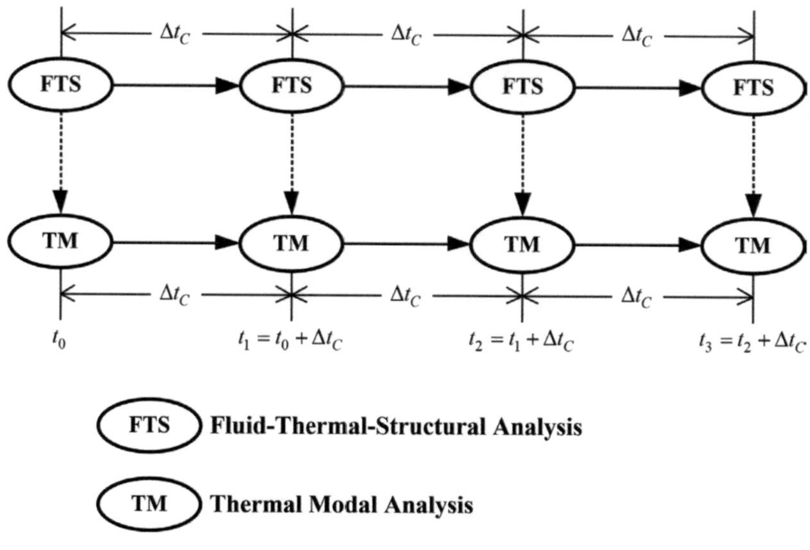

Figure 16. The thermal modal analysis strategy based on multi-physics coupling integration.

5.2.1. Thermal modal analysis along the static trajectory

Set the total cruise flight time to 100 s and the flight environment to be 30 km within standard atmosphere. Calorically perfect gas model is adopted. The thermal modal is analyzed in both the case with zero wing AOA and that with non-zero wing AOA.

Figure 18 presents the first six modal shapes of the hypersonic wing with zero AOA at time t = 100 s. It can be seen that the modal shape of each order at time t = 100 s has local changes to varying degrees compared to the modal shape at initial time. The first four modes generally retain the original modal shapes, while the fifth and the sixth change a lot in their modal shapes with the trend towards bending-torsion coupling that usually leads to vibration of the wing. Therefore, sustained aeroheating has effect on modal shapes of higher order more easily for wings that are fixed-supported at root. As AOA is gradually increased, the aerodynamic forces and thermal load imposed on lower wing surface becomes larger than those imposed on the upper wing surface, which changes the thermodynamic state within the wing and thus influences the natural vibration characteristics of the wing. **Figure 19** presents the first six modal shapes of the hypersonic wing at time t = 100 s under the AOA of 10 deg. It can be seen from the figure that compared with the modal shapes in the zero AOA case, modal shapes of all orders have little change, which indicates that modal shape is not very sensitive to variation of AOA. It is necessary to point out that the analysis above only involves flight time of 100 s. Actually, time of sustained flight of vehicle is an influencing parameter of great importance. The fluid-thermal-structural coupling will finally reach thermodynamic equilibrium. In the process, the modal frequency and modal shape of the wing will continue to change and at the same time the effect of AOA on modal frequency and modal shape will become more prominent.

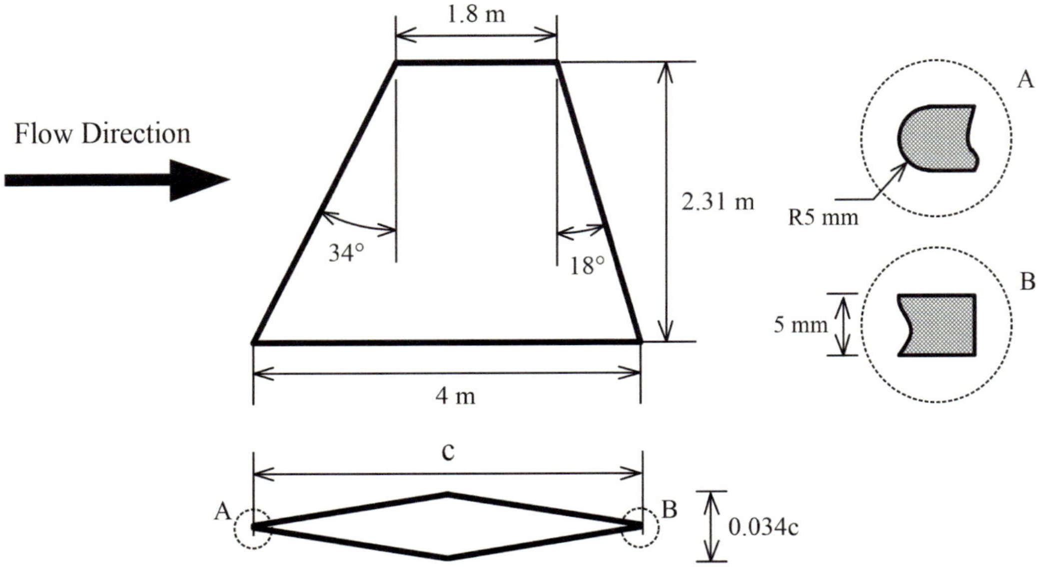

Figure 17. The three-dimensional low-aspect-ratio hypersonic wing.

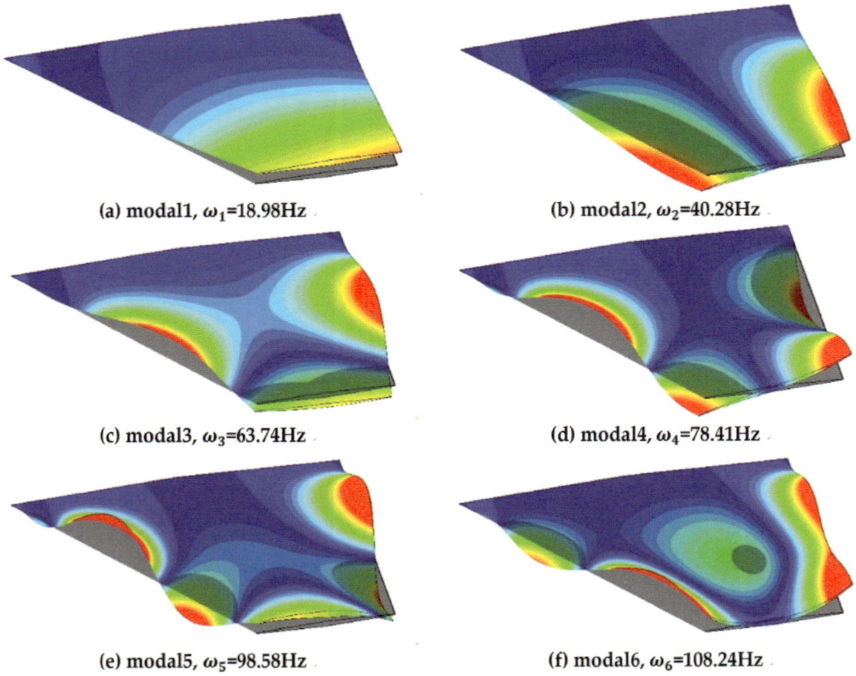

Figure 18. The first six modals of the hypersonic wing with zero AOA at time t = 100 s. (a) modal 1, ω_1 = 18.98 Hz, (b) modal 2, ω_2 = 40.28 Hz, (c) modal 3, ω_3 = 63.74 Hz, (d) modal 4, ω_4 = 78.41 Hz, (e) modal 5, ω_5 = 98.58 Hz, and (f) modal 6, ω_6 = 108.24 Hz.

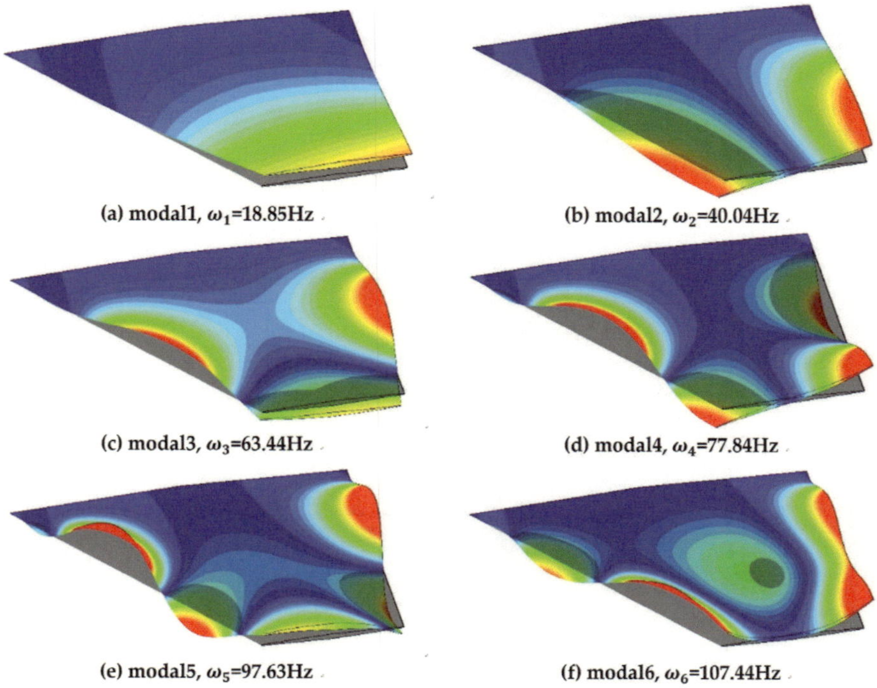

Figure 19. The first six modals of the hypersonic wing at time $t = 100$ s under the AOA of 10 deg. (a) modal 1, $\omega_1 = 18.85$ Hz, (b) modal 2, $\omega_2 = 40.04$ Hz, (c) modal 3, $\omega_3 = 63.44$ Hz, (d) modal 4, $\omega_4 = 77.84$ Hz, (e) modal 5, $\omega_5 = 97.63$ Hz, and (f) modal 6, $\omega_6 = 107.44$ Hz.

5.2.2. *Thermal modal analysis along the dynamic trajectory*

A simple flight trajectory is assumed referring to **Figure 5** to further analyze variation of thermal modal characteristics in complicated flight. **Figure 20** presents that dynamic variation of flight trajectory leads to the variation of the thermodynamic state within the wing as well as its modal frequency. From the initial time to the end of cruise, the modal frequency of each order gradually declines; the decline tends to be gentle and then shows a recovery as the vehicle descends. It turns out that the first modals have little change along the flight trajectory, generally retaining the original modal shapes while major variation takes place in the fifth/sixth-order mode, especially the sixth-order mode in which case bending-torsion coupling tends to occur. If the climbing, cruise and descending phase, especially cruise, last long enough, the modal frequency and modal shape of each order can have more significant variation with dynamic variation of the flight trajectory.

The modal frequencies of each order show a downward trend over time and the decreasing rates vary from mode to mode. Modes of all orders get nearer or farther one another to varying degrees over time, which might impose severe impacts on the natural vibration characteristics of the structure. The results indicate that sustained aeroheating has effect more easily on modal shapes of higher order. Therefore, for hypersonic vehicles with large thin-walled

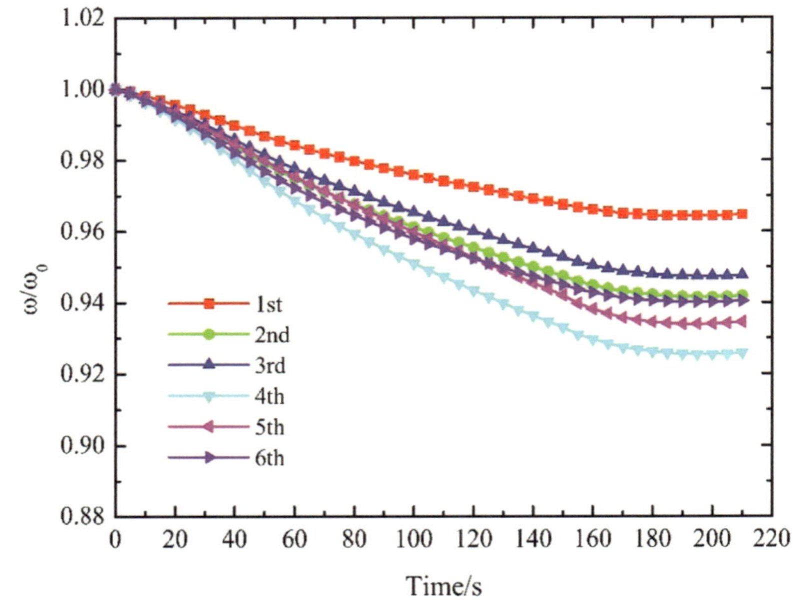

Figure 20. The first six modal frequencies along the flight trajectory.

control surfaces, sustained aeroheating has great effect on the natural vibration characteristics. In general, the strategies and methods for multi-physics field coupling integration analysis developed can effectively predict and analyze the variation of natural vibration characteristics (natural frequency and natural vibration shape), which lays a good foundation for further research on aerothermoelastic problems.

6. Conclusions

By modeling and analysis of hypersonic multi-physics coupling problem, the mathematical and physical description of the various coupling model is established, and then the corresponding coupling analysis strategy is proposed. In the framework of coupling analysis strategy, an integrated analysis program platform HyCCD by integrating hypersonic aerodynamic numerical simulation program and general finite element thermal analysis software is developed to study the relevant problem of hypersonic multi-physics coupling problem. Through these representative studies, the key novel contributions are achieved as follows: (1) studying the coupling mechanism of the multi-physics coupling problems among hypersonic flow, thermal and structure, and hierarchically creating the multi-physics coupling mathematical models and (2) proposing the effective coupling analysis strategies, and synthetically developing a set of high-effective multi-physics coupling integrated analysis method that has engineering applicability. These achievements can provide the theoretical and technical support for the studies on comprehensive performance evaluation and optimization of thermal protection system and the study on aerothermoelastic problem.

Acknowledgements

This study was supported by the National Natural Science Foundation of China (Nos. 11672183, 91641129 and 91441205).

Author details

Fang Chen[1]*, Shengtao Zhang[1,2] and Hong Liu[1]

*Address all correspondence to: fangchen@sjtu.edu.cn

1 Shanghai Jiao Tong University, Shanghai, China

2 AECC Commercial Vehicle Engine Co., Ltd, Shanghai, China

References

[1] Bertin JJ, Cummings RM. Fifty years of hypersonics: Where we've been, where we're going. Progressing in Aerospace Sciences. 2003;**39**(6):511-536

[2] Moses PL, Rausch VL, Nguyen LT, et al. NASA hypersonic flight demonstrators—Overview, status, and future plans. Acta Astronautica. 2004;**55**(3):619-630

[3] McNamara JJ, Friedmann PP. Aeroelastic and aerothermoelastic analysis in hypersonic flow: Past, present, and future. AIAA Journal. 2011;**49**(6):1089-1121

[4] Errera M, Dugeai A, Girodroux-Lavigne P, et al. Multi-physics coupling approaches for aerospace numerical simulations. Aerospace Lab. 2011;(2):1-16

[5] Felippa CA, Park KC, Farhat C. Partitioned analysis of coupled mechanical systems. Computer Methods in Applied Mechanics and Engineering. 2001;**190**(24):3247-3270

[6] Zhang ST, Chen F, Liu H. Time-adaptive, loosely coupled strategy for conjugate heat transfer problems in hypersonic flows. Journal of Thermophysics and Heat Transfer. 2014;**28**(4):635-646

[7] Giles MB. Stability analysis of numerical interface conditions in fluid-structure thermal analysis. International Journal of Numerical Methods in Fluids. 1997;**25**(4):421-436

Chapter 6

Hypersonic Vehicles Profile-Following Based on LQR Design Using Time-Varying Weighting Matrices

Jian Chen, Zihao Xiong, Zixuan Liang, Chen Bai,
Ke Yi and Zhang Ren

Additional information is available at the end of the chapter

http://dx.doi.org/10.5772/intechopen.70659

Abstract

In the process of applying linear quadratic regulator (LQR) to solve aerial vehicle reentry reference trajectory guidance, to obtain better profile-following performance, the parameters of the aerial vehicle system can be used to calculate weighting matrices according to the Bryson principle. However, the traditional method is not applicable to various disturbances in hypersonic vehicles (HSV) which have particular dynamic characteristics. By calculating the weighting matrices constructed based on Bryson principle using time-varying parameters, a novel time-varying LQR design method is proposed to deal with the various disturbances in HSV reentry profile-following. Different from the previous approaches, the current states of the flight system are employed to calculate the parameters in weighting matrices. Simulation results are given to demonstrate that using the proposed approach in this chapter, performance of HSV profile-following can be improved significantly, and stronger robustness against different disturbances can be obtained.

Keywords: hypersonic vehicle, reentry guidance, reference trajectory guidance, linear quadratic regulator, weighting matrix, time-varying

1. Introduction

Hypersonic vehicles (HSV) possess great meaning for aerospace applications, and have important potential values in various fields [1–4]. Reentry guidance of HSV is a critical technology for assuring the vehicle's arrival at a desired destination. The reentry guidance concepts can be described in detail under two general categories [5]; that is, one uses predictive capabilities, and the principal disadvantage is the stringent onboard computer requirements for the fast-time computation; the other one uses a reference trajectory, which provides a simple onboard guidance computation and high reliability of guidance accuracy. The latter has a strong engineering and application value, and has been employed in Apollo entry guidance [6] and shuttle entry guidance [7]. Nevertheless, in reference trajectory reentry guidance, since HSV owns particular characteristics such as strong nonlinear, large flight envelope, complex entry environment, and precise terminal guidance accuracy requirement, it is difficult for HSV to track the nominal profile properly. Many scholars have done continuous research on it [8–12].

For following nominal profile problem in reentry reference trajectory guidance, namely trajectory tracking law, traditional PID control law of shuttle was given in [7]. Roenneke et al. [13] derived a linear control law tracking the drag reference in drag state space to achieve guidance command. In [14], a feedback linearization method for shuttle entry guidance trajectory tracking law to extend its application range was presented. In [15, 16], a feedback tracking law is designed by taking advantage of the linear structure of system dynamics in the energy space to achieve bounded tracking of the flat outputs. These foregoing approaches improved performances of trajectory tracking laws for aerial vehicles. However, they are not applied to HSV which owns particular characteristics. Dukeman [17] proposed a linear trajectory tracking law based on linear quadratic regulator (LQR) by constructing weighting matrices with Bryson principle [18], and the tracking law was very robust with respect to varying initial conditions and worked satisfactorily even for entries from widely different orbits than that of the reference profile. The capacity of the approach in [17] against other reentry process interferences such as aerodynamic parameter error, nevertheless, was relatively poor, and simulations demonstrated that performances for HSV tracking nominal profile under various disturbances depended directly on weighting matrices in LQR. In this study, one focuses on constructing LQR weighting matrices to strengthen robustness of HSV trajectory tracking law.

The weighting matrices Q and R are the most important parameters in LQR optimization and determine the output performances of systems [19]. Trial-and-error method has been employed to construct these matrices, which is simple, but primarily depends on people's experience and intuitive adjustment. In trial-and-error method, elements of weighting matrices must be repeatedly experimented to get a proper value, and is not feasible for application in large scale system. In [18, 20], certain general guidelines were followed to construct weighting matrices simply and normally, but might not lead to satisfactory responses. Connecting closed-loop poles to feedback gains for LQR were presented in [21–24] using pole-assignment approach, which resulted in more accurate responses. However, it was difficult to balance state and control variables and to account for control effectiveness using the approaches in [21–24]. A trade-off between penalties on the state and control inputs for optimization of the cost function was considered in [25], where specified closed-loop eigenvalues were obtained, but the computation normally needed more iterations. Genetic algorithm (GA) can be applied to find a global optimal solution [26–28], and the differential evolution algorithms inspired from GA are efficient evolution strategies for fast optimization technique [29–31]. However, the approaches in [26–31] have little improvement for HSV profile-following performances under different disturbances.

In this study, a novel method to construct weighting matrices with time-varying parameters on the basis of Bryson principle is proposed. This idea employs current flight states to provide flexible and accurate feedback gains in HSV trajectory tracking law under various interferences and errors. Simulations indicate that this approach effectively improves profile-following performance and strengthens the robustness of LQR under different internal and external disturbances.

The rest of this chapter is organized as follows. Section 2 introduces reentry dynamics, LQR, Bryson principle, and their applications in hypersonic vehicle trajectory tracking law. Section 3 analyzes the problem of hypersonic vehicles profile-following. The novel LQR design method with time-varying weighting matrices is presented in Section 4. A numerical simulation is given in Section 5. Finally, Section 6 concludes the whole work.

2. Preliminaries

In this section, the concepts and basic results on reentry dynamics, LQR, Bryson principle, and their applications in hypersonic vehicle trajectory tracking law are introduced, which are the research foundation of the following sections.

2.1. Reentry dynamics

For a lifting reentry vehicle, the common control variables are the bank angle σ, and the angle of attack α. The state variables include the radial distance from the Earth center to the vehicle r, the longitude θ, the latitude φ, the Earth-relative velocity v, the flight path angle γ, and the heading angle ψ. The three-degree-of-freedom point-mass dynamics for the vehicle over a sphere rotating Earth are expressed as [32]:

$$\dot{r} = v \sin \gamma, \tag{1}$$

$$\dot{\theta} = \frac{v \cos \gamma \sin \psi}{r \cos \phi}, \tag{2}$$

$$\dot{\phi} = \frac{v \cos \gamma \cos \psi}{r}, \tag{3}$$

$$\dot{v} = -D - g \sin \gamma + \omega^2 r \cos \phi (\sin \gamma \cos \phi - \cos \gamma \sin \phi \cos \psi), \tag{4}$$

$$\dot{\gamma} = \frac{1}{v}\left[L \cos \sigma - g \cos \gamma + \frac{v^2 \cos \gamma}{r} + 2\omega v \cos \phi \sin \psi + \omega^2 r \cos \phi (\cos \gamma \cos \phi + \sin \gamma \sin \phi \cos \psi)\right], \tag{5}$$

$$\dot{\psi} = \frac{1}{v}\left[\frac{v^2 \cos \gamma \sin \psi \tan \phi}{r} - 2\omega v (\tan \gamma \cos \phi \cos \psi - \sin \phi) + \frac{\omega^2 r}{\cos \gamma} \sin \phi \cos \phi \sin \psi + \frac{L \sin \sigma}{\cos \gamma}\right], \tag{6}$$

where ω is the Earth's self-rotation rate, and g is the gravitational acceleration. L and D are the aerodynamic lift and drag accelerations defined by

$$L = \frac{1}{2m} \rho v^2 C_L S, \; D = \frac{1}{2m} \rho v^2 C_D S, \tag{7}$$

where m is the mass of the vehicle, S is the reference area, C_L is the lift coefficient, and C_D is the drag coefficient. ρ is the atmospheric density expressed as an exponential model [33].

$$\rho = \rho_0 e^{-\beta h}, \tag{8}$$

where ρ_0 is the atmospheric density at sea level, h is the altitude of the vehicle, and β is a constant.

To guide the vehicle from the initial point to the terminal interface with multiple constraints, a reference trajectory is usually optimized offline, and a profile-following law is utilized to track the reference trajectory onboard. In the longitudinal profile-following, the linear quadratic regulator (LQR) law is a good choice [10].

2.2. Linear quadratic regulator

This subsection introduces LQR and Bryson principle. For a linear system, the dynamics can be described by

$$\begin{cases} \dot{x}(t) = A(t)x(t) + B(t)u(t), \\ y(t) = C(t)x(t), \end{cases} \tag{9}$$

where $A(t)$, $B(t)$ and $C(t)$ are system matrices, $x(t) = [x_1(t), x_2(t), \ldots\ldots, x_n(t)]^T$ is the state, and $u(t) = [u_1(t), u_2(t), \ldots\ldots, u_n(t)]^T$ is the control input.

The quadratic performance index required to be minimized can be written as

$$J(t, t_f) = \int_t^{t_f} \left[x^T(\tau)Q(\tau)x(\tau) + u^T(\tau)R(\tau)u(\tau)\right]d\tau, \tag{10}$$

where the weighting matrix $Q(t)$ is symmetrical positive semi-definite and weighting matrix $R(t)$ is symmetrical positive definite. The specific procedure of LQR minimizing quadratic performance index is as follows.

The Riccati equation is given as

$$P(t)A(t) - P(t)B(t)R(t)^{-1}B(t)^T P(t) + Q(t) + A(t)^T P(t) = 0. \tag{11}$$

After getting the solution $P(t)$ corresponding to each time instant t by solving Eq. (11), the feedback gain matrix can be obtained as

$$K(t) = R(t)^{-1}B^T(t)P(t). \tag{12}$$

Based on Eq. (12), the control input can be designed as

$$u(t) = -K(t)x(t). \tag{13}$$

In order to obtain a proper quadratic performance index, elements of weighting matrices must be chosen properly, and Bryson principle can solve this problem effectively.

The basic principle of Bryson principle is to normalize the contributions, and then the states and the control terms may behave effectively within the definition of the quadratic cost function. The normalization is accomplished by using the anticipated maximum values of the individual states and control quantities. The method can be explained as follows.

First, define the weighting matrices $Q(t)$ and $R(t)$ as diagonal matrices, namely:

$$Q(t) = \mathrm{diag}\left[q_1(t), \ldots, q_n(t)\right] \ , \qquad R(t) = \mathrm{diag}[r_1(t), \ldots, r_m(t)]. \tag{14}$$

Then, develop the quadratic index in the following expression.

$$J = \int_t^{t_f} \left(q_1(\tau)x_1(\tau)^2 + \ldots + q_n(\tau)x_n(\tau)^2 + r_1(\tau)u_1(\tau)^2 + \ldots + r_m(\tau)u_m(\tau)^2 \right) d\tau. \tag{15}$$

Determine each maximum value of all the states and control terms.

$$\begin{cases} x_i(\max) \ , \ i = 1, 2, \ldots, n, \\ u_j(\max) \ , \ j = 1, 2, \ldots, m. \end{cases} \tag{16}$$

Normalize all the contributions to 1 with the help of all the maximum values.

$$q_1 x_1(\max)^2 = \ldots = q_n x_n(\max)^2 = r_1 u_1(\max)^2 = \ldots = r_m u_m(\max)^2 = 1. \tag{17}$$

Then, the elements to construct the weighting matrices can be obtained as time-invariant parameters.

$$\begin{cases} q_i = \dfrac{1}{x_i(\max)^2} \ , \ i = 1, 2, \ldots n \ , \\ r_j = \dfrac{1}{u_j(\max)^2} \ , \ j = 1, 2, \ldots m. \end{cases} \tag{18}$$

Through simple calculation, Bryson principle can generate better results in a short time, which minimizes the quadratic index value in a proper scope. Because of that, Bryson principle is widely applied to the selection of weighting matrices in LQR.

2.3. Reentry reference trajectory guidance based on LQR

In reentry reference trajectory guidance, the chief challenge of following the nominal profile lies in generating a proper compensatory signal, and LQR can solve this problem effectively. After generating a feasible reference entry profile containing altitude z, velocity v, flight path angle γ as reference parameters, and bank angle σ_{ref} as guidance reference signal, one can download this profile into the onboard computer. After the vehicle enters the atmosphere, the deviations between nominal profile and the actual real-time data can be obtained by navigation facilities. The deviations contain altitude error z_δ, velocity error v_δ and flight path angle error γ_δ.

Denote the compensatory bank angle by σ_δ. In order to minimize the deviations and keep the aerial vehicle tracking nominal profile properly, the optimal feedback gain of guidance compensatory signal σ_δ can be calculated by LQR in the following algorithm.

Algorithm 1. The actual guidance signal in trajectory tracking law based on LQR can be determined in the following procedure.

Step 1. Based on perturbation theory, one establishes the linear equations of motion by taking the deviations as state parameters.

$$\begin{cases} \delta x'(t) = A(t)\delta x(t) + B(t)\delta u(t), \\ y(t) = C(t)\delta x(t). \end{cases} \tag{19}$$

where $\delta x(t) = [z_\delta(t), v_\delta(t), \gamma_\delta(t)]^T$ and $\delta u(t) = \sigma_\delta(t)$.

Step 2. Construct the quadratic performance index as follows:

$$J(t, t_f) = \int_t^{t_f} \left[\delta x^T(\tau) Q \delta x(\tau) + \delta u(\tau) R \delta u(\tau)\right] d\tau. \tag{20}$$

Step 3. The weighting matrices Q and R in Eq. (20) are determined by Bryson principle. Since altitude z and velocity v are main factors in profile-following, q_3, which is the weighting element of path angle γ, can be ignored. Using Eq. (18), the other elements of weighting matrices can be obtained as

$$q_1 = \frac{1}{z_{\delta\max}{}^2}, \quad q_2 = \frac{1}{v_{\delta\max}{}^2}, \quad r_1 = \frac{1}{\sigma_{\delta\max}{}^2}, \tag{21}$$

where $z_{\delta\max}$ and $v_{\delta\max}$ are anticipated maximum deviations between the actual profile and the nominal profile, and $\sigma_{\delta\max}$ is the maximum allowable modification of guidance signal σ. Based on Eq. (21), one can get the weighting matrices:

$$Q = diag\left[q_1, q_2\right], R = r_1. \tag{22}$$

Step 4. In order to minimize the index J in Eq. (20), one calculates the Riccati Eqs. (11) and (12) to obtain the optimal feedback gain $K(t)$. Then, the compensatory signal can be obtained as

$$\delta u(t) = -K(t)\delta x(t). \tag{23}$$

Step 5. The actual guidance signal $\sigma(t)$ which consists of guidance reference signal $u(t)$ and guidance compensatory signal $\delta u(t)$ can be obtained. It can be shown that

$$\sigma(t) = u(t) + \delta u(t) = \sigma_{ref} - K(t)\delta x(t). \tag{24}$$

It has been verified in [17, 34] that using Algorithm 1, the aerial vehicle performs well in tracking reference trajectory.

3. Problem statement

In this section, the problem of HSV profile-following using trajectory tracking law based on LQR in Section 2 is presented.

The traditional reference guidance is not suitable for hypersonic vehicles because of its particular characteristics, including strong nonlinear, large flight envelope and complex entry environment. In the process of entry flight, it is difficult to constrain the deviation between real profile and nominal profile into a proper scope. Furthermore, strict terminal accuracy requirement demands that hypersonic vehicles track nominal profile precisely, that is, deviations in the terminal stage must be smaller.

Consequently, in the reference profile-following of HSV based on LQR, new problems occur in the selection of weighting matrices. In the initial flight stage, it is assumed that the deviations of altitude and velocity are $z_{\delta 0}$ and $v_{\delta 0}$, respectively, which are chosen to be

$$\begin{cases} z_{\delta 0} = 3\,\mathrm{km}, \\ v_{\delta 0} = 200\,\mathrm{m/s}. \end{cases} \tag{25}$$

Let $z_{\delta 1}$ and $v_{\delta 1}$ be the anticipated maximum deviations accuracy in the terminal stage, expressed as

$$\begin{cases} z_{\delta 1} = 0.5\ \mathrm{km}, \\ v_{\delta 1} = 20\mathrm{m/s}. \end{cases} \tag{26}$$

Substituting $z_{\delta 0}$, $v_{\delta 0}$, $z_{\delta 1}$ and $v_{\delta 1}$ into Eq. (21), one can get the weighting matrix Q_0 and Q_1 as

$$Q_0 = \begin{bmatrix} \frac{1}{{z_{\delta 0}}^2} & 0 \\ 0 & \frac{1}{{v_{\delta 0}}^2} \end{bmatrix}, \quad Q_1 = \begin{bmatrix} \frac{1}{{z_{\delta 1}}^2} & 0 \\ 0 & \frac{1}{{v_{\delta 1}}^2} \end{bmatrix}. \tag{27}$$

From Eqs. (25) and (26), one sees that $z_{\delta 0}$ and $v_{\delta 0}$ are bigger than $z_{\delta 1}$ and $v_{\delta 1}$, respectively. The weighting matrix Q_0, which is determined by $z_{\delta 0}$ and $v_{\delta 0}$, can effectively eliminate the large initial stage deviations between the real and nominal profiles. Nevertheless, the capacity of Q_0 for resisting disturbance in the process of flight is not strong enough to satisfy the terminal accuracy requirement. On the contrary, the weighting matrix Q_1 constructed by $z_{\delta 1}$ and $v_{\delta 1}$ can eliminate the disturbance in the process of flight effectively. However, facing the existence of large deviations in the initial entry flight, it is difficult to keep HSV tracking the nominal profile properly, which will further influence the terminal accuracy.

Therefore, compared with Q_0, the weighting matrix Q_1 is not applicable to initial deviations, and has good robustness to deal with disturbance in the process of entry flight. The

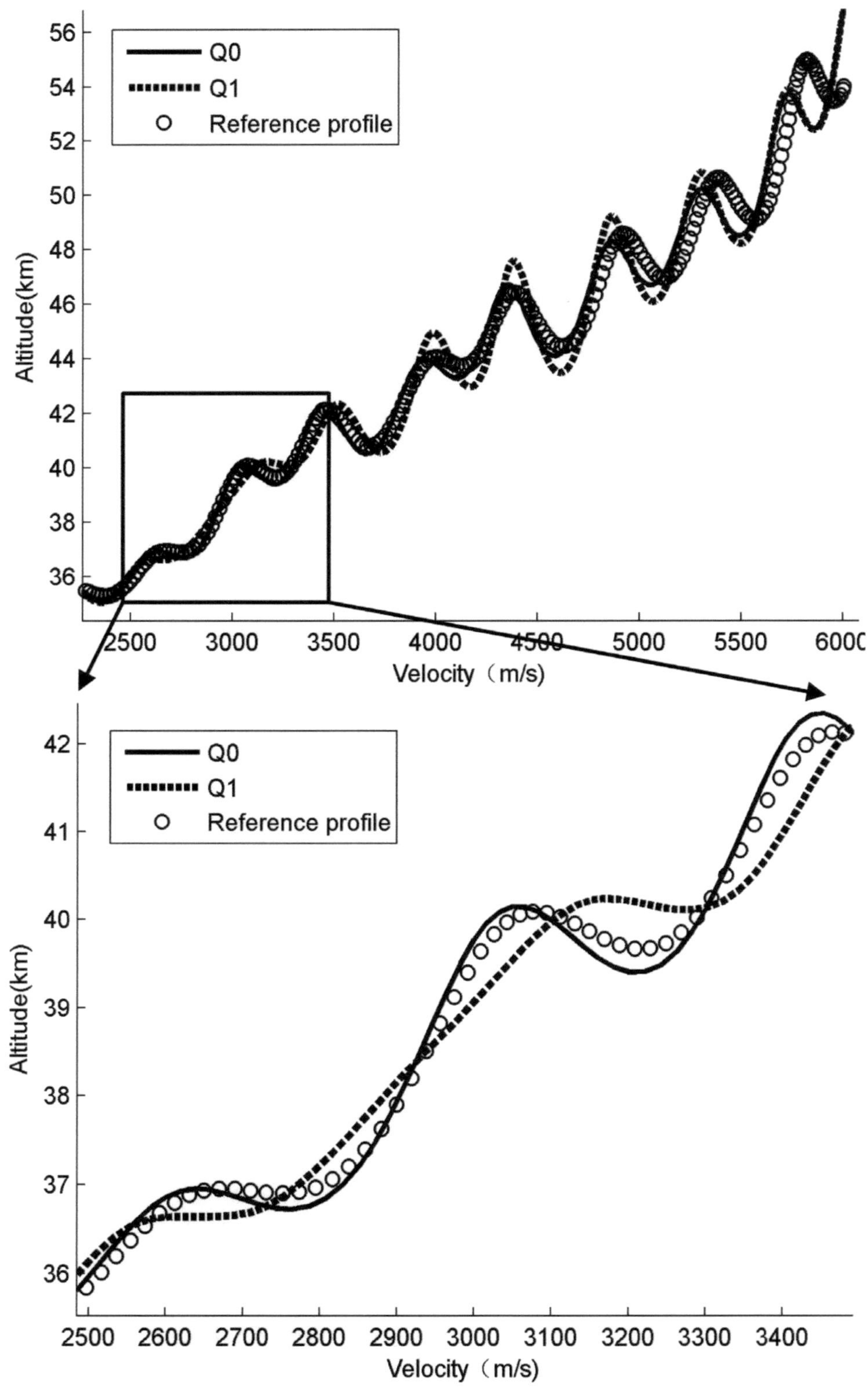

Figure 1. The profile-following of HSV entry guidance with initial 3 km altitude deviation by Q_0 and Q_1.

http://dx.doi.org/10.5772/intechopen.70659

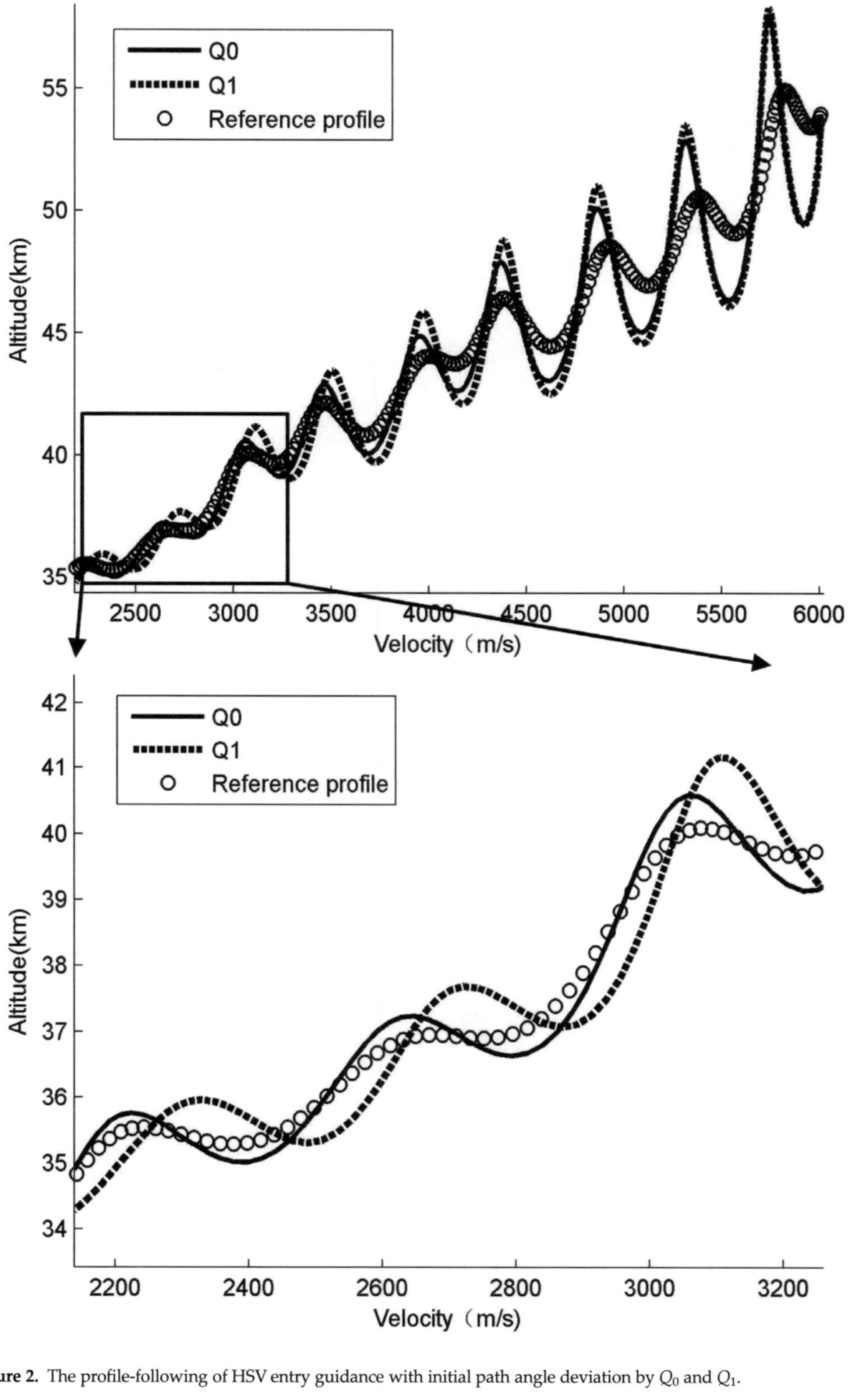

Figure 2. The profile-following of HSV entry guidance with initial path angle deviation by Q_0 and Q_1.

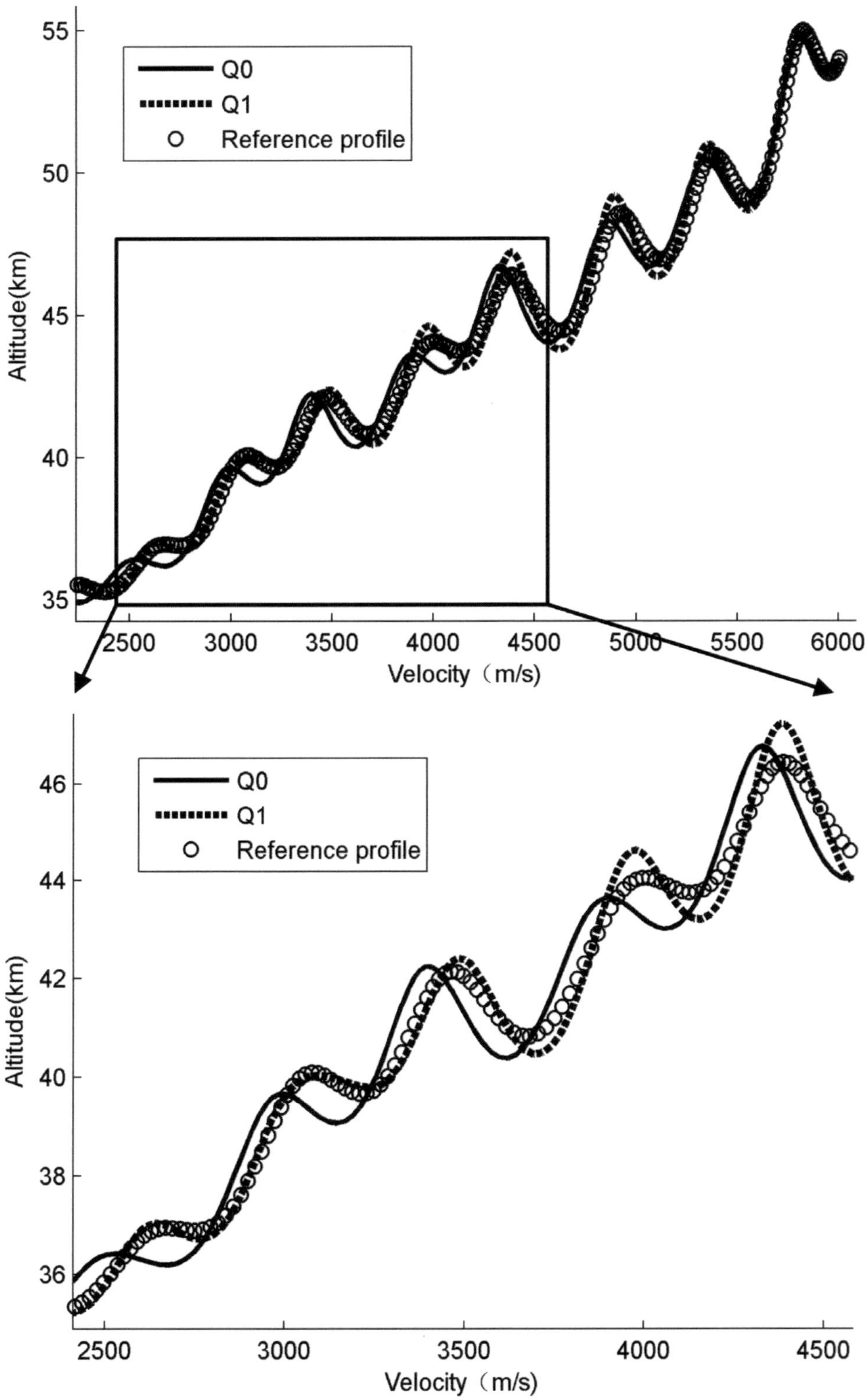

Figure 3. The profile-following of HSV entry guidance with 20% aerodynamic parameter error by Q_0 and Q_1.

simulations for hypersonic vehicles profile-following with Q_0 and Q_1 under different disturbances are shown in **Figures 1–3**. The flight profiles are expressed in altitude and velocity plane.

The reference profile described in this study is similar to the shuttle entry reference profile. The initial altitude of simulation is 55 km, and the initial velocity is 6 km/s.

Figure 1 shows that hypersonic vehicle tracks nominal profile with initial altitude deviation of positive 3 km, where circle line, solid line, dashed line indicate nominal profile, actual profile with Q_0, actual profile with Q_1, respectively. From **Figure 1**, one sees that the performance of Q_0 tracking nominal profile is better than Q_1. **Figures 2** and **3** are in respect to HSV profile-following with initial deviation of path angle and process disturbance of positive 20% aerodynamic parameter error. It can be seen that the performance of Q_0 tracking nominal profile is better than Q_1 in **Figure 2**, and Q_1 is better than Q_0 in **Figure 3**. Based on **Figures 1–3**, it can be obtained that the LQR with weighting matrices constructed by Bryson principle hasn't strong robustness to different disturbances in HSV profile-following.

In order to solve above problem, it is required that LQR cannot only minimize the initial deviations, but also enhance the capability that resists the process disturbance effectually. Therefore, it is necessary to develop an algorithm to determine a proper weighting matrix in LQR. With the help of Bryson principle, an approach to determine the weighting matrix in LQR with current flight states is proposed in the following section.

4. LQR with time-varying weighting matrices

In this section, first, the flow chart of HSV profile-following is presented. Then, LQR design method using time-varying weighting matrix for HSV reentry trajectory tracking law is derived.

Based on LQR, here is the flow chart of HSV tracking reference profile shown as the solid lines in **Figure 4**.

The work flow of HSV profile-following is explained as follows:

Comparing the actual flight profile with the reference profile, one can get the state deviations containing z_δ and v_δ. With these deviations, the compensatory signal u_δ can be calculated by

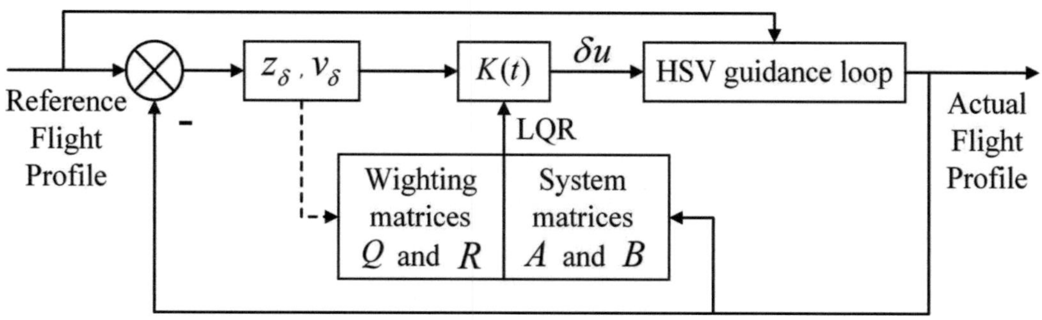

Figure 4. The flow chart of HSV profile-following based on LQR.

multiplying feedback gain K. Then one can input the compensatory signal and the reference guidance signal into HSV guidance loop. In this way, actual flight profile of the next step is obtained. The calculation of the feedback gain K by LQR involves four matrices. As shown in the **Figure 4**, the construction of system matrices A and B needs actual state parameters. Weighting matrices Q and R need to be determined and downloaded into the onboard computer before starting entry guidance of HSV.

Instead of obtaining the specific elements in traditional method, the LQR design method using time-varying weighting matrix substitutes the flight state deviations z_δ and v_δ into the calculation of Q. The main idea of this method can be explained as the dashed line in **Figure 4**. With the help of Bryson principle, the calculation of elements in weighting matrix Q involves two parameters $z_{\delta\max}$ and $v_{\delta\max}$. These two parameters represent maximal allowable deviations in altitude and velocity between actual and reference profiles, respectively. In the time-varying optimization method, one can make a comparison between the actual real-time profile and the relevant reference profile, and get the current deviations $z_\delta(t)$ and $v_\delta(t)$. Then substitute them into $z_{\delta\max}$ and $v_{\delta\max}$, that is,

$$z_{\delta\max} = z_\delta(t) \ , v_{\delta\max} = v_\delta(t). \tag{28}$$

Substituting $z_{\delta\max}$ and $v_{\delta\max}$ into Eq. (21), the weighting matrix Q can be obtained. The following algorithm is proposed to determine the actual guidance signal $\sigma(t)$ with time-varying weighting matrix in LQR.

Algorithm 2. The actual guidance signal in trajectory tracking law based on LQR using time-varying weighting matrix can be designed in the following procedure.

Step 1 Measure the actual current flight profile which contains altitude z and velocity v. Compare them with the relevant reference altitude z_{ref} and velocity v_{ref}, the current deviations z_δ and v_δ can be obtained, respectively.

Step 2 Substitute z, v, and γ into system, and calculate the linear system matrices $A(t)$ and $B(t)$.

Step 3 Construct the weighting matrices $Q(t)$ and $R(t)$ by substituting z_δ, v_δ and maximal allowable adjustment of guidance signal $\sigma_{\delta\max}$ into Eqs. (28), (21) and (22).

Step 4 Calculate the feedback gain $K(t)$ by $A(t)$, $B(t)$, $Q(t)$, $R(t)$ in Eqs. (11) and (12).

Step 5 The compensatory guidance signal δu can be calculated by $K(t)$ in Eq. (23). Then one can get the actual guidance signal in Eq. (24), namely bank angle $\sigma(t)$.

5. Simulation results

In this section, a numerical simulation is given to demonstrate the effectiveness of the proposed method in the previous section.

Q_0 and Q_1 are defined in Section 3, and the time-varying weighting matrix is denoted by Q. The simulations for hypersonic vehicle following reference profile with Q are shown in **Figures 5–7**.

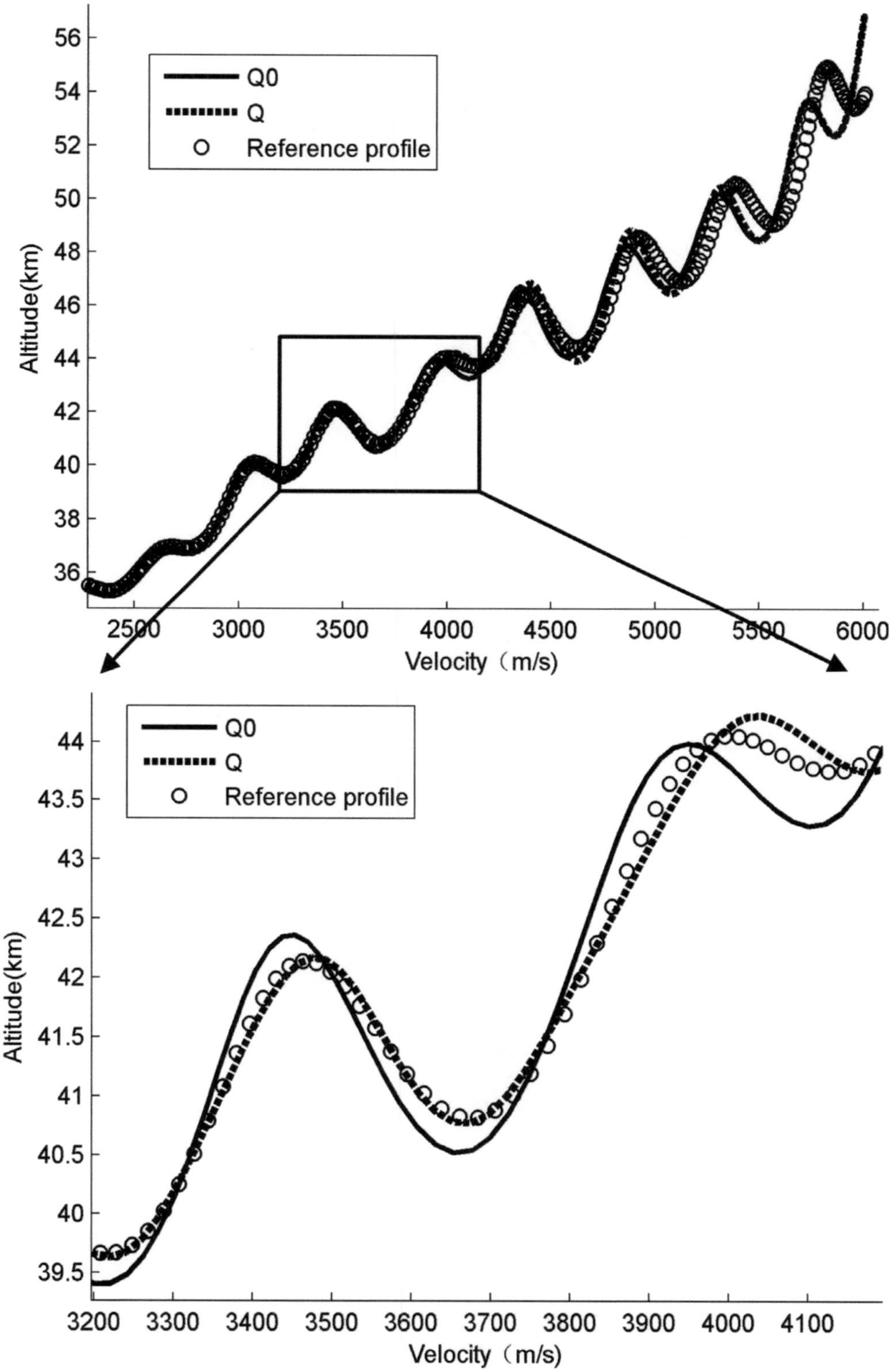

Figure 5. The profile-following of HSV entry guidance with initial altitude deviation by Q_0 and Q.

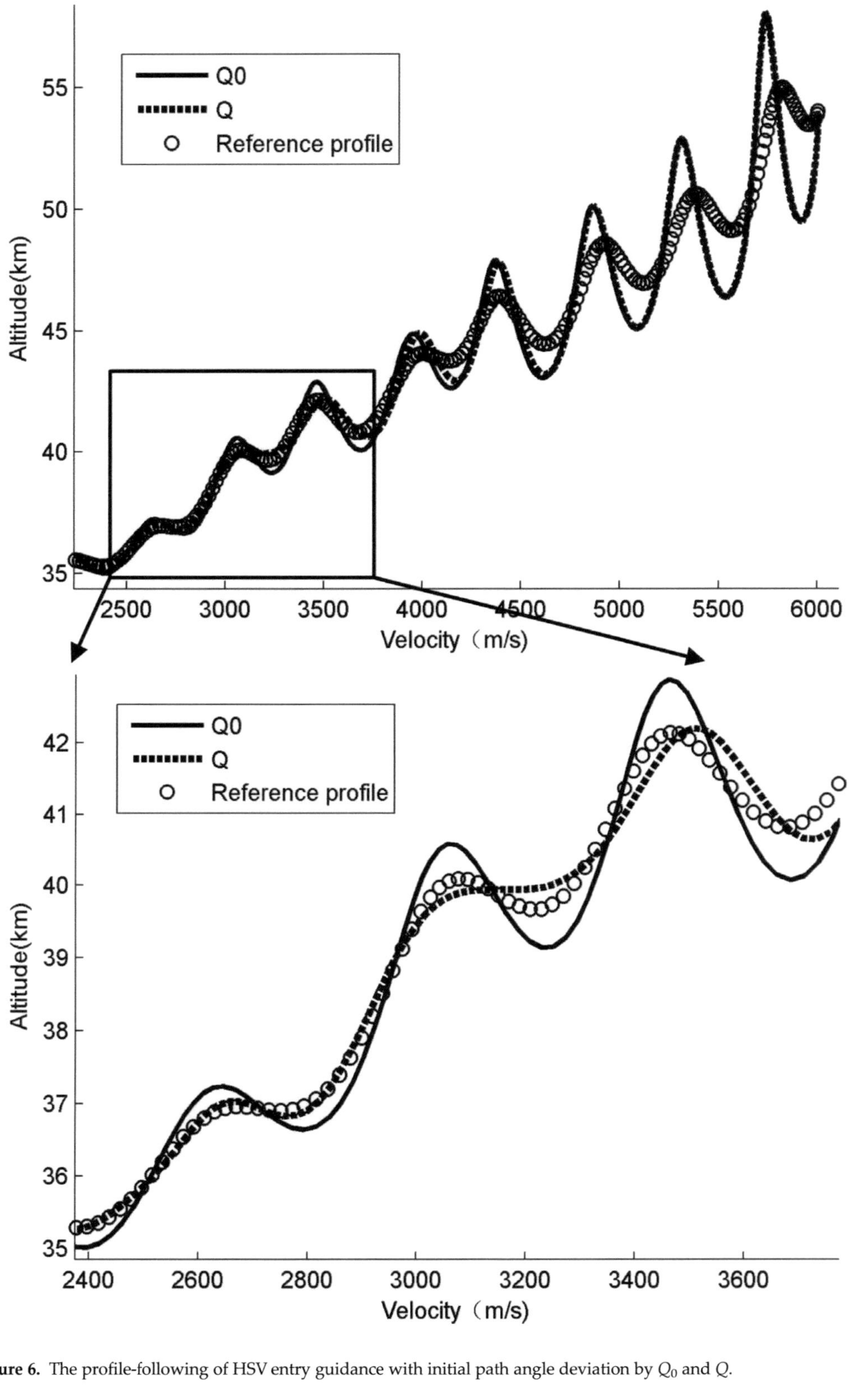

Figure 6. The profile-following of HSV entry guidance with initial path angle deviation by Q_0 and Q.

http://dx.doi.org/10.5772/intechopen.70659

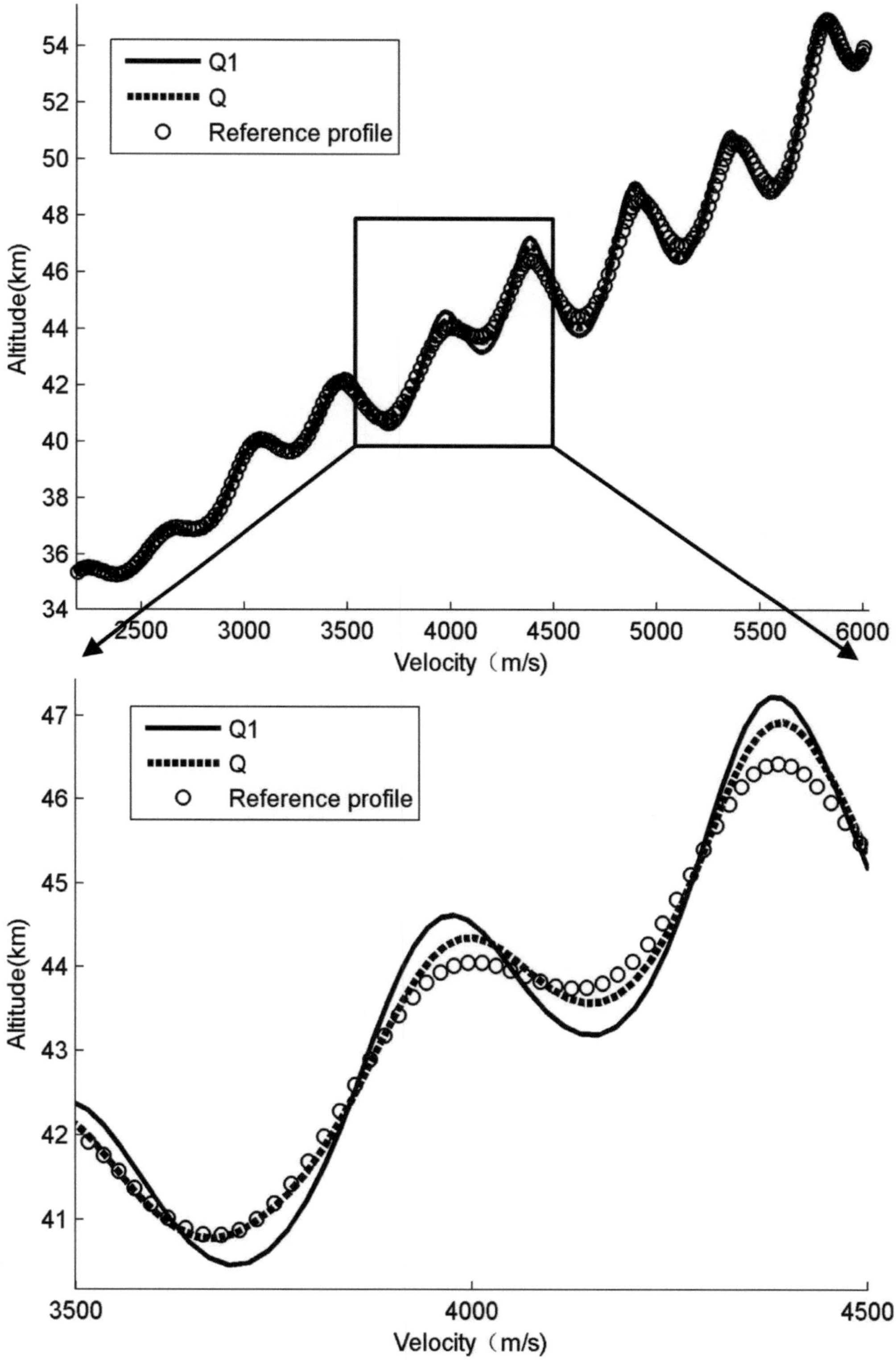

Figure 7. The profile-following of HSV entry guidance with 20% aerodynamic parameter error by Q_1 and Q.

For initial deviations of altitude and path angle, it's clear that the profile-following performances of Q_0 are better than Q_1 from **Figures 1** and **2**. Consequently, for the same deviations, **Figures 5** and **6** choose Q_0 to compare with Q. Since Q_1 performs better than Q_0 in **Figure 3** under the aerodynamic parameter error, one can choose Q_1 to compare with Q under the same disturbance in **Figure 7**.

From **Figures 5–7**, it can be shown that the time-varying matrix Q has better performance than Q_0 and Q_1 in the application of LQR on HSV following reference profile.

6. Conclusions

Because of complex entry environment and particular dynamic characteristics, such as strong nonlinear and large flight envelope, LQR with weighting matrices constructed by traditional Bryson principle was not suitable for HSV profile-following in reentry reference trajectory guidance. On the basis of Bryson principle, this chapter proposed time-varying matrices constructed by current flight states. The capability of HSV tracking nominal profile using LQR with time-varying weighting matrices was significantly improved. From simulations, it could be clearly shown that the LQR designed by presented method was more robust than traditional way to different initial deviations and disturbances in the process of reentry flight.

Acknowledgements

This work is supported by National Key R&D Program of China under Grant Nos. 2017YFD0701000 and 2016YFD0200700, Chinese Universities Scientific Fund under Grant Nos. 2017QC139 and 2017GX001, and National Natural Science Foundation of China under Grant No. 61333011.

Author details

Jian Chen[1]*, Zihao Xiong[2], Zixuan Liang[2], Chen Bai[3], Ke Yi[4] and Zhang Ren[2]

*Address all correspondence to: chenjian@buaa.edu.cn; jchen@cau.edu.cn

1 College of Engineering, China Agricultural University, Beijing, China

2 Beihang University, Beijing, China

3 Beijing Institute of Aerospace System Engineering, Beijing, China

4 Systems Engineering Research Institute, China State Shipbuilding Corporation, Beijing, China

References

[1] Hallion R. The history of hypersonics:Or, 'back to the future: Again and again'. In: 43rd AIAA Aerospace Sciences Meeting and Exhibit; 10–13 January 2005. Reno, Nevada: American Institute of Aeronautics and Astronautics; 2005

[2] Anderson GY. An outlook on hypersonic flight. In: 23rd Joint Propulsion Conference; 29–02 July 1987. San Diego, CA, USA: American Institute of Aeronautics and Astronautics; 1987

[3] McClinton CR, Rausch VL, Shaw RJ, Metha U, Naftel C. Hyper-X: Foundation for future hypersonic launch vehicles. Acta Astronautica. 2005;**57**(2–8):614-622

[4] Voland RT, Huebner LD, McClinton CR. X-43A hypersonic vehicle technology development. Acta Astronautica. 2006;**59**(1–5):181-191

[5] Wingrove RC. Survey of atmosphere re-entry guidance and control methods. AIAA Journal. 1963;**1**(9):2019-2029

[6] Graves CA, Harpold JC. Apollo experience report: Mission planning for Apollo entry. In: NASA Technical Note; 1972. p. NASA-TN-D-6725, MSC-S-305

[7] Harpold JC, Graves CA. Shuttle entry guidance. In: American Astronautical Society, Anniversary Conference; Houston, USA. Oct. 30-Nov. 2, 1978. American Astronautical Society, United States. pp. 239–268

[8] Huang W, Ma L, Wang ZG, Pourkashanian M, Ingham DB, Luo SB, Lei J. A parametric study on the aerodynamic characteristics of a hypersonic waverider vehicle. Acta Astronautica. 2011;**69**(3–4):135-140

[9] Block RF, Gessler Jr GF, Panter WC, Konsewicz RK. The challenges of hypersonic-vehicle guidance, navigation, and control. In: Space Programs and Technologies Conference; 25-27 September 1990; Huntsville, AL: American Institute of Aeronautics and Astronautics; 1990

[10] Shen ZJ, Lu P. Onboard generation of three-dimensional constrained entry trajectories. Journal of Guidance, Control, and Dynamics. 2003;**26**(1):111-121

[11] Johnson EN, Calise AJ, Curry MD, Mease KD, Corban JE. Adaptive guidance and control for autonomous hypersonic vehicles. Journal of Guidance, Control, and Dynamics. 2006;**29**(3):725-737

[12] Lu P, Forbes S, Baldwin M. Gliding guidance of high L/D hypersonic vehicles. In: AIAA Guidance, Navigation, and Control Conference; 19-22 August 2013; Boston, MA; American Institute of Aeronautics and Astronautics; 2013

[13] Roenneke AJ, Markl A. Reentry control to a drag vs. energy profile. In: AIAA Guidance, Navigation and Control Conference; 9-11 August 1993; Monterey, CA; American Institute of Aeronautics and Astronautics; 1993

[14] Mease KD, Kremer JP. Shuttle entry guidance revisited using nonlinear geometric methods. Journal of Guidance, Control and Dynamics. 1994;**17**(6):1350-1356

[15] Bharadwaj S, Rao A, Mease KD. Entry trajectory tracking law via feedback linearization. Journal of Guidance, Control and Dynamics. 1998;**21**(5):726-732

[16] Bharadwaj S, Rao A, Mease KD. Tracking law for a new entry guidance concept. In: 22nd Atmospheric Flight Mechanics Conference; 11–13 August 1997 New Orleans, LA, U.S.A: American Institute of Aeronautics and Astronautics; 1997

[17] Dukeman GA. Profile-following entry guidance using linear quadratic regulator theory. In: AIAA Guidance, Navigation and Control Conference and Exhibit; 05-08 August 2002; Monterey, California; American Institute of Aeronautics and Astronautics; 2002

[18] Bryson AE, Ho YC, Siouris GM. Applied optimal control: Optimization, estimation, and control. IEEE Transactions on Systems, Man and Cybernetics. 1979;**9**(6):366-367

[19] Zhang HY, Qing W. Optimal Control Theory and Application. Beijing: Higher Education Press; 2006

[20] Stein G. Generalized quadratic weights for asymptotic regulator properties. IEEE Transactions on Automatic Control. 1979;**24**(4):559-566

[21] Andry AN, Shapiro EY, Chung JC. Eigenstructure assignment for linear systems. IEEE Transactions on Aerospace and Electronic Systems. 1983;**19**(5):711-729

[22] Fahmy M, O'reilly J. On eigenstructure assignment in linear multivariable systems. IEEE Transactions on Automatic Control. 1982;**27**(3):690-693

[23] Chu KWE. A pole-assignment algorithm for linear state feedback. Systems and Control Letters. 1986;**7**(4):289-299

[24] Lee WC, Lyu YY, Hsu TY, Wei CC. Using the Taguchi methods to study the balance control of a rotary inverted pendulum. In: 2014 International Conference on Advanced Robotics and Intelligent Systems; 06–08 June 2014; Taipei, China. IEEE, United States. 2014. pp. 171–175

[25] Luo J, Lan CE. Determination of weighting matrices of a linear quadratic regulator. Journal of Guidance, Control and Dynamics. 1995;**18**(6):1462-1463

[26] Goldberg DE. Genetic Algorithms in Search, Optimization and Machine Learning. Reading, Massachusetts: Addison-Wesley; 1989

[27] Davies R, Clarke T. A parallel implementation of the genetic algorithm applied to the flight control problem. In: IEE Colloquium on High Performance Computing for Advanced Control; 8–8 December 1994. London, UK: IET; 1994. pp. 6/1-6/3

[28] Molazadeh VR, Banazadeh A, Shafieenejad I. Design of the LQR controller and observer with fuzzy logic GA and GA-PSO algorithm for triple an inverted pendulum and cart system. In: International Conference on Advanced Mechatronic Systems; 10-12 August 2014. Kumamoto, Japan: IEEE; 2014. pp. 295–300

[29] Storn R, Price K. Differential evolution—a simple and efficient heuristic for global optimization over continuous spaces. Journal of Global Optimization. 1997;**11**(4):341-359

[30] Liouane H, Chiha I, Douik A, Messaoud H. Probabilistic differential evolution for optimal design of LQR weighting matrices. In: International Conference on Computational Intelligence for Measurement Systems and Applications; 2-4 July 2012. Tianjin, China: IEEE; 2012. pp. 18–23

[31] Tijani IB, Akmeliawati R, Abdullateef AI. Control of an inverted pendulum using MODE-based optimized LQR controller. In: 8th IEEE Conference on Industrial Electronics and Applications; 19-21 June 2013. Melbourne, Australia: IEEE; 2013. pp. 1759–1764

[32] Vinh NX, Busemann A, Culp RD. Hypersonic and Planetary Entry Flight Mechanics. Ann Arbor: University of Michigan Press; 1980

[33] Seiff A. Atmospheres of earth, mars and venus, as defined by entry probe experiments. Journal of Spacecraft and Rockets. 1991;**28**(3):265-275

[34] Cavallo A, Ferrara F. Atmospheric re-entry control for low lift/drag vehicles. Journal of Guidance, Control and Dynamics. 1996;**19**(1):47-53

Chapter 7

Porous Ceramic Matrix Phase Change Composites for Thermal Control Purposes of Hypersonic Vehicle

Xiangfa Zhou

Additional information is available at the end of the chapter

http://dx.doi.org/10.5772/intechopen.70863

Abstract

Thermal control systems and heat insulation materials are required for a range of hypersonic vehicles ranging from ballistic reentry to hypersonic cruise vehicles, both within Earth's atmosphere and non-Earth atmospheres. The combined thermodynamic/heat transfer relations of the phase change materials (PCMs) in silica nanoporous materials are developed to obtain mass, thickness, and temperature excursion as functions of percentage area of PCM under given maximum energy and thermal flux. The studies show that PCMs are one of the most preferred methods to thermal control applications that can effectively delay or modify the temperature rise of the surface of the aircrafts subjected to high thermal flux. This chapter also introduces the preparations of porous ceramic matrix phase change composite, putting PCMs to use in the internal thermal control materials for the hypersonic vehicles. Porous ceramic matrix serves as the supporting material, which provides structural strength and prevents the leakage of melted PCMs, and PCMs act as thermal absorb material limiting the temperature abruptly rising of the aircrafts. The structural pore properties of the silica matrix with different molar ratios of ethanol (EtOH)/tetraethoxysilane (TEOS) are investigated to determine suitable porous matrices for PCM. To adjust the pore structure of porous silica matrices with different molar ratios of EtOH and TEOS for PCM infiltration is mainly discussed. Furthermore, numerical and experimental studies are proposed to predict and investigate the thermal absorption characteristics of porous silica infiltrated with PCM for thermal control applications.

Keywords: ceramic matrix composite (CMC), phase change materials (PCMs), thermal control systems, preparation, heat transfer modeling

1. Introduction

Controlling the temperature and/or heat transfer is of crucial importance in numerous technical processes and natural occurrences, including cooling of electronics and telecommunications equipment in aerospace applications. Phase change materials (PCMs), which store excess heat generated from the work components and release it reversibly when needed, provide a smart approach for more efficient temperature management and thermal control utilization. Latent heat absorption phenomena associated with melting of a suitable PCM may be effectively used to delay or modify the temperature rise of the surface subjected to high heat flux. Thermal management utilizing solid-liquid PCMs is one of the most interesting passive thermal management techniques due to its simplicity and reliability. Solid-liquid PCMs including salt hydrates, paraffin waxes, certain hydrocarbons, and metal alloys are often used for thermal protection applications.

Most practical applications for hypersonic vehicles require PCMs to have high density of latent heat and appropriate phase change point. In this criterion, certain hydrocarbons are very promising for temperature controlling applications for its comparative high fusion latent heats, suitable melting temperature, and chemical stability [1]. However, the applications of the kinds of solid-liquid PCMs are largely limited, as leaking of the liquid phase occurs above the phase change point. Thus, accommodation of the PCMs in appropriate host materials is necessary to prevent the leakage of the liquid phase for temperature controlling applications using the solid-liquid phase change. The leakage is usually circumvented by introducing shape stabilization support [2, 3]. Devoting to develop composites with a high PCM load and heat storage density, researchers have been recently interested in silica nanoporous ceramics with high specific area as a shape stabilization matrix or a supporting structure materials as PCMs' skeleton for avoiding leakage of solid-liquid transition [4, 5]. Silica aerogels have been used extensively in many engineering applications. They are sol-gel-derived porous inorganic materials recognized for their low density (as low as 0.01–0.02 g/cm^3) [6], high porosity (>90%) [7], and high specific surface area (600–1000 m^2/g) [8]. These structural properties result in a few thermal conductivities and high optical transparencies [9]. However, monolithic silica aerogels are very fragile and transparent to thermal radiation at high temperature, and these two defects restraint the application range of silica aerogels. Fortunately, researchers have improved both the mechanic strength property and high transparency using reinforced fibers [10] and pacifiers doped in the aerogels [11]. The composite silica aerogels are called silica nanoporous materials, and they have the similar space structure and have superior properties such as higher mechanic strength properties at high temperature [12, 13]. Thus, they have better applicability than the monolithic aerogels. The silica nanoporous materials have broad applications or application prospect in the fields which have strict limit of space or weight, for example, in aeronautics and aerospace [14].

The concept of thermal design using PCM has been well established through various studies in recent decades and is widely used in space applications [15]. To apply PCMs as alternating heat storage and discharge in space applications, the material should accommodate the special needs on the thermophysical properties that include conformance of phase change temperature to the design limits, high latent heat, and low density difference between liquid and solid phases. Structurally and thermally, the synthesis of the support materials for a PCM must be also considered. The supporting materials assure the seepage-proof for the molten PCMs and can withstand some imposed mechanical loads structurally. The supporting materials must be also not degraded the system performance thermally. To design PCM thermal protection system properly, the thermodynamic conservation relations and heat transfer must be considerate.

In this chapter, the combined thermodynamic/heat transfer relations of the PCM in silica nanoporous materials are developed to obtain mass, thickness, and temperature excursion as functions of percentage area of PCM under given maximum energy and thermal flux. And then, the preparation and characterization of form-stable PCM/porous silica ceramic composite are reported. The paraffin acts as thermal absorb material, and porous silica serves as the supporting material, which provides structural strength and prevents the leakage of melted PCM. To adjust the pore structure of porous silica matrices with different molar ratios of EtOH and TEOS for PCM infiltration is mainly discussed. At last, numerical and experimental studies are proposed to predict and investigate the thermal absorption characteristics of porous silica infiltrated with PCM for thermal control applications. The numerical simulation was performed using a volume-averaging technique, and a finite volume modeling (FVM) was used to discretize the heat diffusion equation. The phase change process was modeled using the enthalpy-porosity method.

2. Theory: modeling of combined thermodynamic/heat transfer

To develop the combined thermodynamic/heat mass transfer relations for the PCM/silica nanocomposite, two additional assumptions are made in the development of the relations: firstly, contact resistance between the PCM and the silica nano-materials is assumed negligible, and, secondly, three-dimensional heat transfer effects are neglected, based on the assumption that the skeleton of silica nano-materials is so closely spaced that these effects are negligible.

When the component temperature rise for a particular application exceeds the maximum operational temperature of component, the supporting material raises the equivalent thermal conductivity of the PCM by providing low thermal-resistance paths through the PCM and reduces the temperature gradient necessary to dissipate the imposed cold-plate heat load. The reduction in temperature gradient reduces the temperature excursion of the component. Consider the thermal protection system shown in **Figure 1**. An electrical component is thermally protected by the PCM/silica nanocomposite system. The component internally generates energy cyclically. For cyclical operation, three equations can be derived for the system based on the model.

2.1. Conservation of energy

After the maximum energy that must be stored by the PCM/silica nanocomposite, E_{Max}, is determined, the following heat balance will hold.

$$E_{Max} = \rho_P \cdot A_P \cdot t \cdot h_f + \left[\rho_S \cdot A_S \cdot C_S + \rho_P \cdot A_P \cdot C_P\right] \cdot \frac{t}{2} \cdot (T_{Max} - T) \tag{1}$$

where ρ_P, C_P, and A_P are density, specific heat, and area of PCM, respectively; t is the thickness of the nanocomposite; ρ_S, A_S, and C_S are the density, specific heat, and area of silica, respectively;

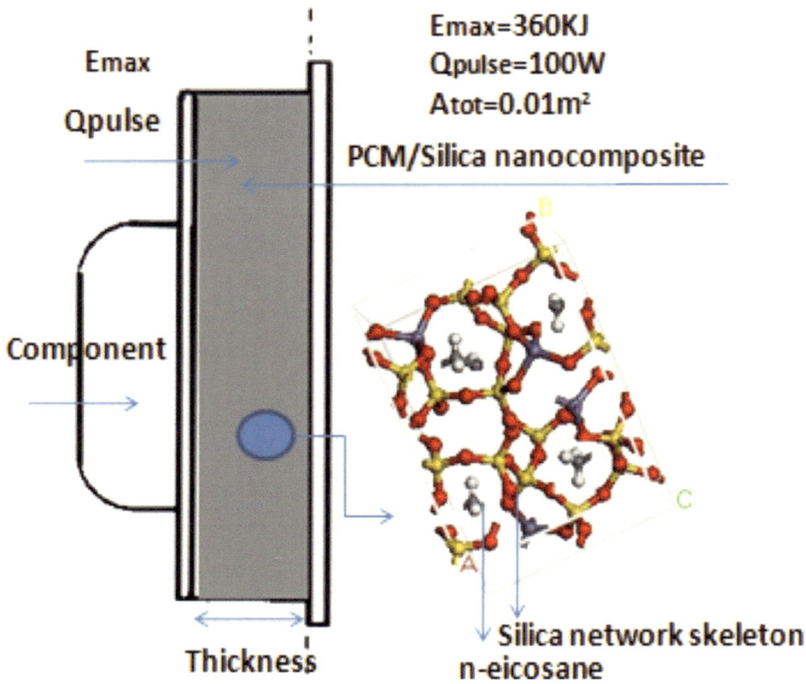

Figure 1. PCM thermal control system and model of PCM/silica nanocomposite.

and $T_{Max} - T$ is the temperature excursion of the component. This equation treats both the energy stored through latent heat of fusion and sensible heat stored within the liquid PCM and the silica nano-materials.

2.2. Conservation of mass

The mass balance shown below will hold:

$$M_{tot} = \left(\rho_P \cdot A_P + \rho_S \cdot A_S\right) \cdot t \tag{2}$$

where M_{tot} is the total mass of the nanocomposite.

2.3. Temperature range constraints

The equation establishes a relation between the total conductivity, K_{tot}; the total area, A_{tot}; the thickness, t; and the temperature excursion $T_{Max} - T$

$$Q_{pulse} = \frac{K_{tot} \cdot A_{tot} \cdot (T_{Max} - T)}{t} \tag{3}$$

2.4. Additive relations

For parallel conductance, the total equivalent conductance can be found from the following equation

$$K_{tot} \cdot A_{tot} = k_P \cdot A_P + k_S \cdot A_S \tag{4}$$

This relation neglects three-dimensional effects and contact resistances. The total area is the sum of the cross-sectional areas of the PCM and silica nano-materials.

Thus,

$$A_{tot} = A_P + A_S \tag{5}$$

2.5. Thermodynamic/heat transfer relations

2.5.1. Thermophysical properties

Thermophysical properties are shown in **Table 1**. As shown in **Table 1**, it is concluded that n-eicosane's phase transition at a moderate temperature makes it a candidate phase change materials or PCM which can be used to store thermal energy and control temperature.

2.5.2. Structural and modeling

Silica open-network structure supported by fibers distributed uniformly within. The reinforced silica open-network structure with high porosity and high surface area will benefit for liquid n-eicosane impregnation. The large surface area and low density of porous materials will enhance the shape stabilization capability and thus maximize the surface area per unit volume of PCM. When used in a spacecraft, the materials thermal control system must be small and light. The n-eicosane impregnates in the open network of silica nano-materials as an integrated thermal control system. Among the composite structure, the structural silica nano-materials function not only as reinforced skeleton of the composite but also as heat-conducting materials to transfer heat from the component to the low thermal conductive PCM. The thermal design meets the requirement of total PCM service from alternate melting and freezing during the whole period of mission.

According to this composite structure, the system combining PCM and silica nano-materials is suggested as a new type of thermal control device for the generated heat electronic component.

Name	n-Eicosane ($C_{20}H_{42}$)	Silica nanoporous material
Melting point (°C)	36.7*	–
Latent heat of fusion per unit mass (J/kg × 10^{-3})	247*	–
Specific heat (J/kg K)	2210 at 308 K**	549**
Thermal conductivity (W/m K)	0.15**	0.37**
Density (kg/m^3)	856 at 308 K***	280***

*Measured by DSC.
**Measured by LFA 1000 Laserflash (thermal conductivity/diffusivity).
***Calculated by experiment.

Table 1. Thermophysical properties of n-eicosane and silica nanoporous material.

The PCM thermal control system and the cross section of designed structure, where the PCM is embedded in silica nano-materials, is depicted in **Figure 1**. The cross section of the n-eicosane/silica nanocomposite is an area with 100 mm × 100 mm. The performance of the thermal control device is investigated by numerical analysis. The main purpose of the study is to obtain the functions and relations between the mass, thicknesses, temperature excursion, and percentage area of n-eicosane through combined thermodynamic/heat transfer analysis, which raises the minimum temperature and lowers the maximum temperature. According to the thermal cycle period, composite PCMs accumulate heat and uniformly redistribute it. Thus, the phase change temperature of PCM is the target of temperature control, and it must remain within the maximum/minimum operating temperature range of the electronic component. Furthermore, it must be chemically stable for silica nano-materials, which is the material of PCM support, and there must be a small density difference between solid and liquid phases.

2.5.3. Thermodynamic/heat transfer relations

It is assumed that the component heat is generated periodically according to the system. In detail, there is 3.6×10^2 kJ the maximum energy stored by the PCM/silica nanocomposite, Emax, is determined and then a warming up period of 100 W heating, Q-pulse, during the whole cycle. To obtain a quantitative idea of the functional relationships between the variables described above for a given application, thermodynamic/heat transfer was conducted using the given power and energy requirement, using a given area, and using the thermophysical data of the PCM and silica nano-materials, as described in **Table 1**. These simultaneous Eqs. (1)–(5) can be solved to yield the excursion temperature, T_{Max}-T; the total mass, $M(t)$; and the thickness, t, as functions of PCM area percentage, A_{pcm}/A_{tot}. A MATLAB program was coded to solve the five equations and yields the parametric data discussed above.

The hyperbolic function curve about the relationship between the area percentage of PCM and the excursion temperature, as shown in **Figure 2**, illustrates one interesting fact. The intersections of the curve with the $A_{pcm}/A_{tot} = 0$ vertical line represent the temperature excursion for only a solid silica nano-material heat sink, which means the temperature excursion for a nanocomposite without n-eicosane. The temperature excursion is highest at this condition and decreases drastically with small additions of n-eicosane until a smaller point is reached around 50% n-eicosane. Similarly, the intersections of the curve with the $A_{pcm}/A_{tot} = 1.0$ vertical line represent the temperature excursion for the whole fusion heats of PCM. At this condition, the temperature excursion reaches the smallest value, showing the inferiority of a heat sink of solid silica nano-materials compared to PCM/silica nanocomposite. But the opposite is also true (black curve as shown in **Figure 2**), showing the PCM better controlling ability and repeatability of temperature when intensively change.

The curves about the relationship between the area percentage of PCM and the total mass and thickness, as shown in **Figure 3**, illustrate a monotonic decrease in mass and thickness quantities with addition of PCM. The intersections of the curve with the $A_{pcm}/A_{tot} = 0$ vertical line represent the mass or thickness for only a solid silica nano-materials. The total mass or thickness is most for this condition, and addition of n-eicosane causes a monotonic decrease in both quantities. However, the temperature excursion is highest at this condition when compared to **Figure 2**. Similarly, the intersections of the curve with the $A_{pcm}/A_{tot} = 1.0$ vertical

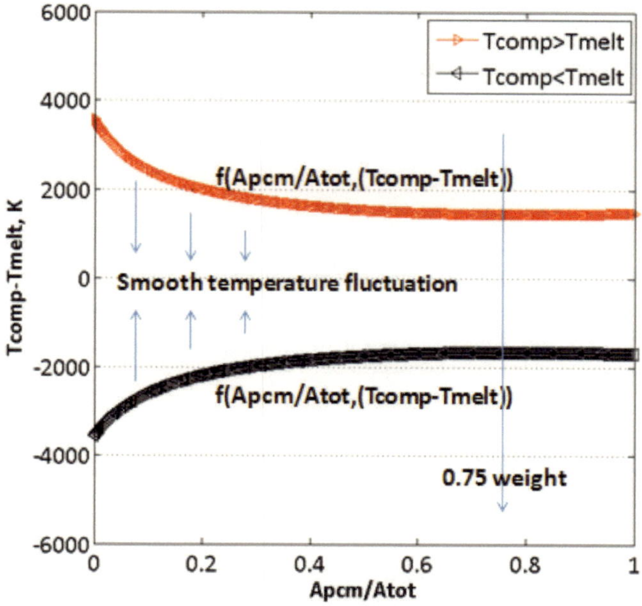

Figure 2. Relationship between the area percentage of PCM and the excursion of temperature.

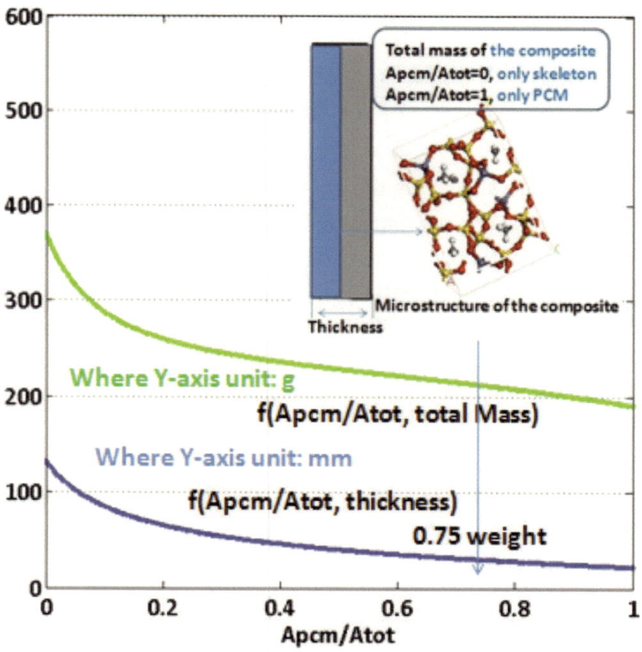

Figure 3. Relationship between the area percentage of PCM and the total mass and thickness.

line represent the mass or thickness for PCM. At this condition, the mass and thickness reach their minimum values, showing also the PCM better controlling the ability of temperature with less mass or volume quantities.

2.6. Enlightening

We performed combined thermodynamic/heat transfer analysis to obtain the total mass, thickness, and temperature excursion as functions of percentage area of PCM under given maximum energy and thermal flux based on the composite structural model and the measured thermophysical data. The relationship between the area percentage of PCM and the excursion temperature is the hyperbolic function, showing the nanocomposite better temperature control management. The relationship between the area percentage of PCM and total mass and thickness shows the nanocomposite better temperature control without much mass or volume quantities. These results are attributed to the strong interaction between the n-eicosane and the silica skeleton, which exhibits novel temperature management and energy utilization.

3. Preparation: characterization and properties of paraffin/porous silica ceramic composites

A form-stable composite of paraffin/porous silica composite was prepared by sol-gel method and melt infiltration. The properties of the composite will be characterized by means of DSC, SEM, and FTIR. Furthermore, the mass fractions of the paraffin in silica matrices prepared by different molar ratios of EtOH/TEOS were recorded to determine the relation between the mass fraction of paraffin and the silica matrices.

3.1. Materials and methods

3.1.1. PCM

The technical grade paraffin (n-eicosane: the chemical formula $C_{20}H_{42}$) used in this study was supplied by Nanyang Chemicals Company in China. Thermophysical data of the PCM are given in **Table 1**. The melting temperature ranges, and fusion heats of the PCMs were measured by a DSC instrument (NETZSCH STA449C) with heating rate at 10°C/min in the temperature range of 20–80°C.

3.1.2. Preparation of n-eicosane/silica nanocomposite

The preparation process of n-eicosane/silica nanoporous composite is shown in **Figure 4**. Firstly, the support materials—porous silica for the PCM—were synthesized by a two-step sol-gel process. Secondly, reinforced fibers (main compositions: SiO_2 > 99.95%) were exploited as structural reinforcement matrix.

The fibrous materials were uniformly pre-shaped in a mold and sol-gel mixtures and closely infiltrated the fibers by pouring them into the mold. Then, the mold includes silica sol-gel mixtures with fiber material uniformly distributed within which were supercritically dried in an autoclave above the critical temperature and critical pressure after gelation and aging in ethanol. The illustration of supercritical fluid drying equipment is shown in **Figure 4**. The final bulk silica nanoporous materials are obtained.

http://dx.doi.org/10.5772/intechopen.70863

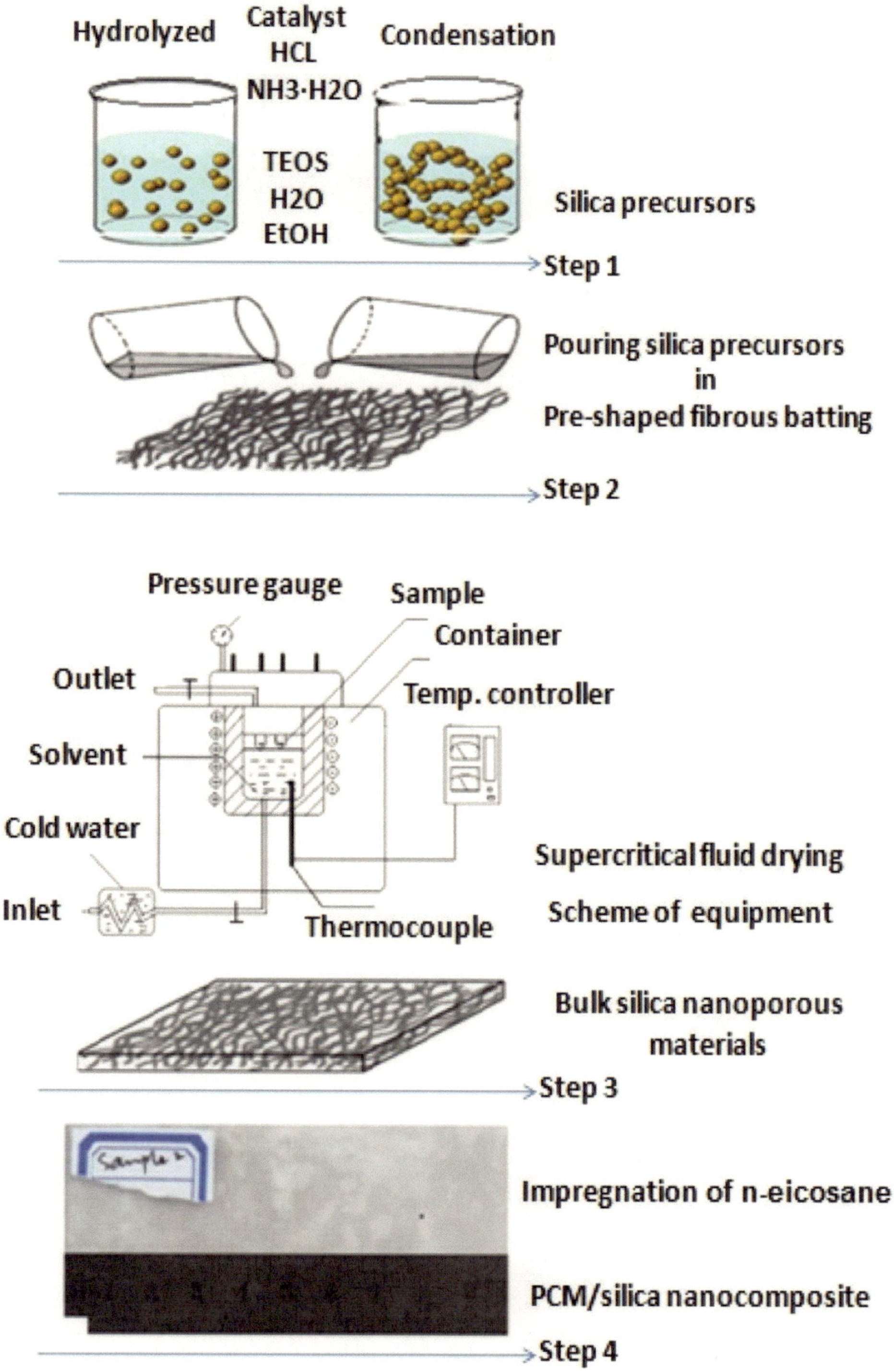

Figure 4. Experimental preparation process of n-eicosane/silica nano-composite.

3.1.3. Pore structure modification of silica matrix infiltrated with paraffin

The molar ratios of EtOH and TEOS were designed as the reactants to prepare the porous silica. The final silica showed a different pore structure with the change of EtOH/TEOS. The mass fraction of the melted paraffin infiltrated into the porous silica was closely related to the pore structure of silica. The mass fractions of the paraffin in silica matrices prepared by different molar ratios of EtOH/TEOS were recorded to determine the relation between the mass fraction of paraffin and the silica matrices.

3.2. Characterization and properties

3.2.1. Pore structure modification of porous silica matrices

The adsorption-desorption measurements were performed by ASAP 2000 micromeritic apparatus. Analysis of the surface areas of the porous silica samples was conducted by Brunauer-Emmett-Teller (BET) method [16]. The method also determines the external/mesoporous surface area and extends the analysis of adsorbed gas from the lower P/P_0 range of Langmuir mono-layer and BET multilayer adsorption to the higher P/P_0 range. The pore size is also determined using the Barrett-Joyner-Halenda (BJH) method [17, 18].

The N_2 adsorption-desorption isotherms of the silica samples are given in **Figure 5**. It shows that the sorption isotherms of the samples with the moral ratios of the EtOH/TEOS (E) = 2, E = 3, and E = 5 belong to Type IV mesoporous structures according to IUPC classification, nitrogen absorption of which was linearly increased with the relative pressure until 0.92 and then increased abruptly and saturated at near saturated pressure during absorption process. During desorption, nitrogen was dissociated, and the cumulative absorbed volume decreased greatly in the region of 0.90–0.95. For E = 10 porous silica, the region of great desorption change shifted to 0.97–1. It was shifted to higher region for E = 20 porous silica, and the steep slopes may be associated with capillary condensation in larger pores (>50 nm) and classified as Type II according to the Kelvin equation and IUPC classification. N_2 sorption data indicates that the E = 2, E = 3, and E = 5 silica are comprised of small-sized pores compared to those of

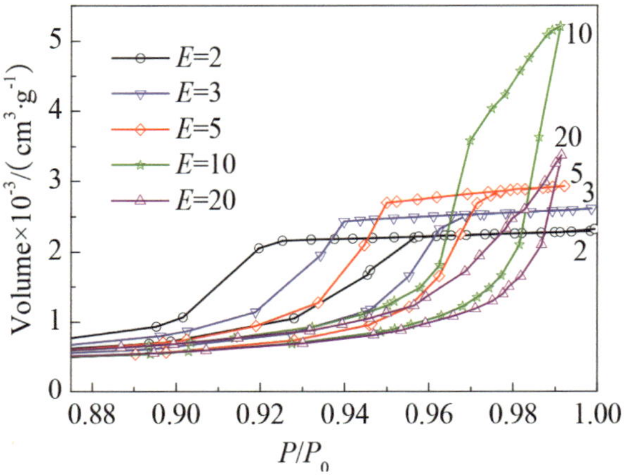

Figure 5. The N_2 adsorption-desorption isotherms of the silica samples.

the E = 10 and E = 20 silica. Pore structural characteristics of the samples obtained from the analysis of N_2 sorption data were shown in **Table 2**.

Figure 6 shows the pore size distributions of samples with different molar ratios of EtOH/TEOS measured by nitrogen desorption method (BJH method). The network formations of silica are known to be affected by water/Si, catalyst, pH, temperature, and ethanol/Si [19]. Results show that the EtOH/TEOS molar ratio causes the difference of silica network structures because the greater ethanol content retards the gelation time and results in the network structures with larger pores when the other basic compositions hold constant.

Microphotographs are taken from the fracture surface of the silica samples using JSM-6360LV scanning electron microscope. SEM of the microstructures of samples with the three kinds of EtOH/TEOS molar ratios of 2, 10, and 20 is given in **Figure 7**. The pore size increases when the molar ratios of the EtOH/TEOS increase by comparing with the samples with the three kinds of molar ratios of the EtOH/TEOS; the pore size of the sample with the EtOH/TEOS molar ratio of 20 is bigger than the other two types of the EtOH/TEOS molar ratios of 2 and 10. Several

EtOH/TEOS	Smulti ($m^2 \cdot g^{-1}$)	Ssingl ($m^2 \cdot g^{-1}$)	C	V ($cm^3 \cdot g^{-1}$)	Davera (nm)	DBJH (nm)
2	618.3	583	44.31	3.594	23.3	23.4
3	624.8	577.7	88.42	4.042	25.9	32.4
5	565.4	518.9	83.86	4.534	32.1	38.5
10	606.5	552.8	94.22	8.051	53.1	59.5
20	596.7	573	91.55	5.219	56	60.7

Note: Smulti, multiple point BET values; Ssingl, mono-point BET values; V, pore volume; Davera, mean pore diameter; DBJH, pore diameter by BJH method.

Table 2. Pore characteristics of the samples with various molar ratios of EtOH/TEOS (E) obtained from the analysis of N_2 sorption data.

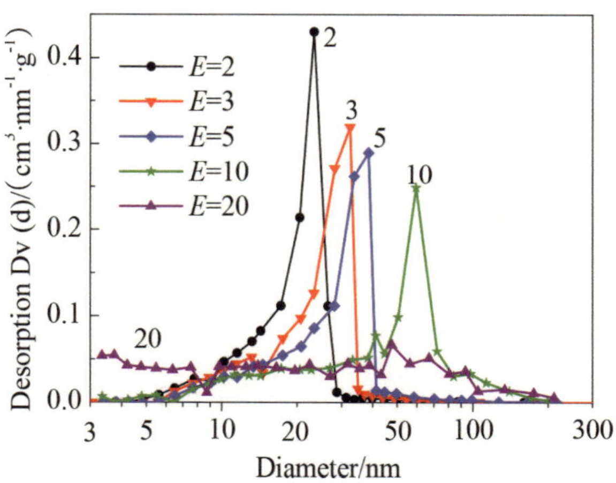

Figure 6. The pore size distributions of samples with different molar ratios of EtOH/TEOS measured by nitrogen desorption method (BET).

kinds of form-stable paraffin/porous silica composites were successfully prepared in our recent study. Using the silica matrices synthesized from the EtOH/TEOS ratio of 10, the PCM showed better thermal absorption characteristics, which is suitable for the fields of requiring cooling at high temperatures such as aircraft electronics and spacecraft devices.

3.2.2. The mass fractions of the paraffin

The relation between the mass fraction of the paraffin infiltrated the three kinds of molar ratios of EtOH/TEOS (E = 2, E = 10, E = 20) silica matrices and the infiltration time was plotted in **Figure 8**. The mass fraction of the paraffin in porous silica increases with the infiltration time, and the E = 10 and E = 20 silica matrix composites reached to 75% at 180 s, while E = 2 silica matrix composite reached to 68%, and then they all increase a little with further time. It indicates that the maximum mass fractions of the paraffin as PCM in E = 10 and E = 20 silica matrices are over than one of the E = 2 silica matrices and the bigger mass fraction of the

Figure 7. SEM of the microstructures of samples with the three kinds of EtOH/TEOS molar ratios of 2, 10, and 20.

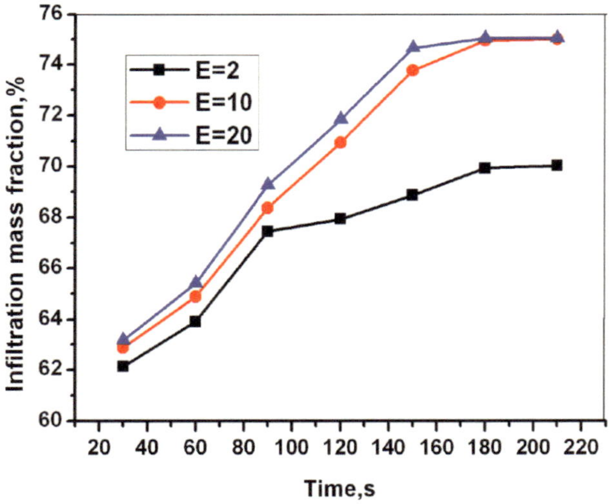

Figure 8. The relation between the mass fractions of the paraffin infiltrated the three kinds of molar ratios of EtOH/TEOS (E = 2, E = 10, E = 20) silica matrices and the infiltration time.

paraffin meant considerable latent heat energy storage potential. The pore size of porous silica becomes larger with the increase of the EtOH/TEOS molar ratios; when the molar ratios of EtOH/TEOS are 10 and 20, the average pore size of the synthesized silica are 53.1 and 56.0 nm, respectively. The maximum mass percentage of paraffin as PCM in the E = 10 and E = 20 silica matrices reached to 75 wt% and over than those of the E = 2 silica matrix. When the molar ratios of EtOH/TEOS are 10 and 20, the synthesized silica matrices are suitable to serve as supporting materials of the PCM.

3.2.3. *Structure analysis*

The morphologies of silica nanoporous materials and PCM/silica composite were observed with a field emission scanning electron microscope (FE-SEM: JSM-6700F, JEOL, Japan, with a thin Pt-Pd coating). **Figure 9** compares the SEM of the microstructures of porous silica and the composite. The microstructure of the porous silica (**Figure 9a**) is porous and connected each other, while the microstructure of the composite (**Figure 9b**) shows that the paraffin was dispersed uniformly into the porous network of silica ceramics, which used as supporting material and provided a mechanical strength to the PCM. The maximum mass fraction of paraffin dispersed into the composites was measured as 75%, and there was no leakage of the paraffin from the surface of the composite up to this mass fraction even over its melting temperature.

3.2.4. *Chemical properties*

FTIR spectra of the paraffin and the silica and the paraffin/porous silica composite are shown in **Figure 10**. In FTIR spectra of the paraffin, the peaks at the wave numbers of 2935 and 2860 cm^{-1} are caused by stretching vibration of C–H, peaks at around 1500 cm^{-1} belong to the deformation vibration of $-CH_2$ and $-CH_3$, and the peak at 750 cm^{-1} represents the rocking vibration of $-CH_2$. In FTIR spectra of the silica, the peaks at 1012 and 810 cm^{-1} are caused by bending vibration of Si–O, and an intense band at the wave number of 1090 cm^{-1} is typical of

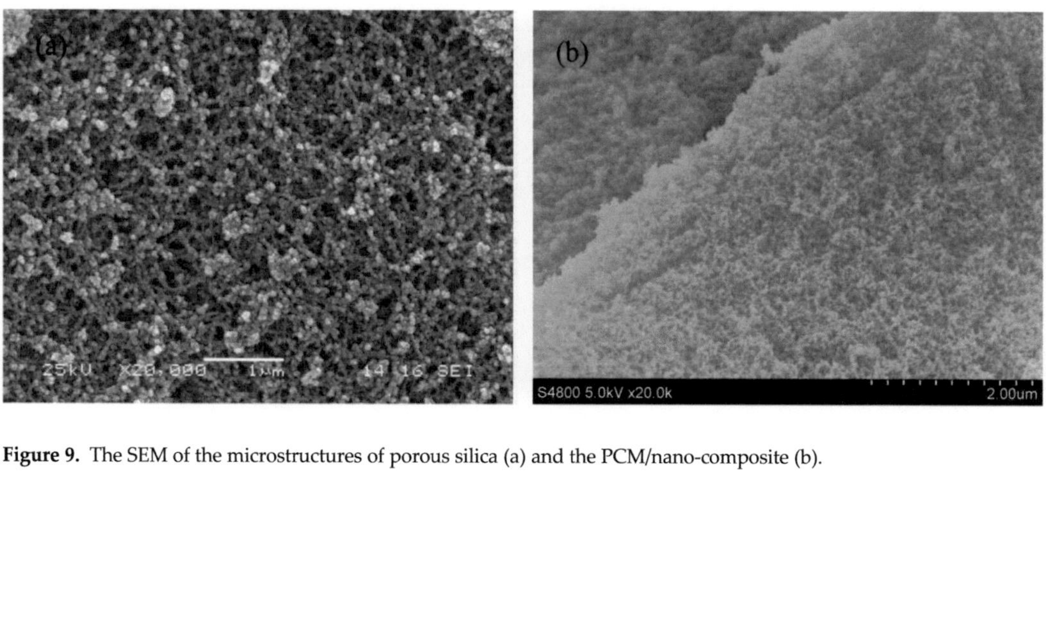

Figure 9. The SEM of the microstructures of porous silica (a) and the PCM/nano-composite (b).

the bending vibration of Si–O. And, the peak at 3449 cm^{-1} represents the stretching vibration of functional group of Si–OH. For the composite, the peaks at the wave numbers of 3500, 2935, 2860, 1500, 1090, and 750 cm^{-1} have corresponding vibration, and no significant new peaks were observed. The FTIR spectra illustrate that the composite is just a physical combination of silica ceramics and paraffin.

3.2.5. Thermal properties

The DSC curves of the paraffin and the composite are shown in **Figure 11**. From **Figure 1a**, the latent heat of the paraffin is 182.22 kJ/kg (T_m = 56.8°C). TG curve shows that weight of paraffin hardly changes, which indicates the paraffin used as PCM has good thermal chemical stability. **Figure 11b** indicates that heat storage capacity of the composite happened at melting point

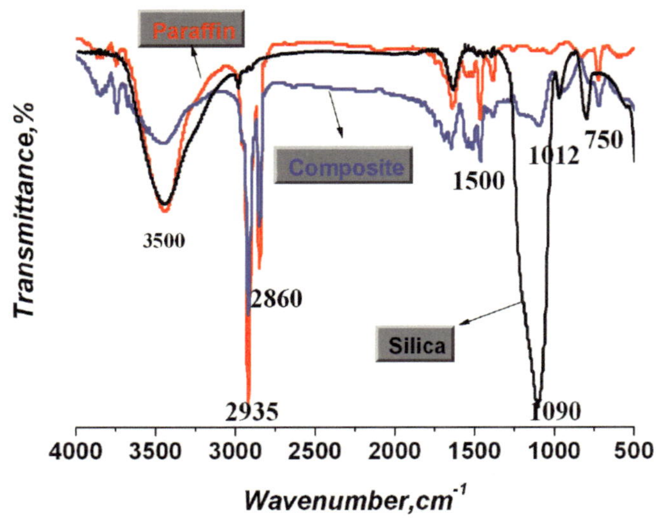

Figure 10. FTIR spectra of the paraffin and the silica and the paraffin/porous silica composite.

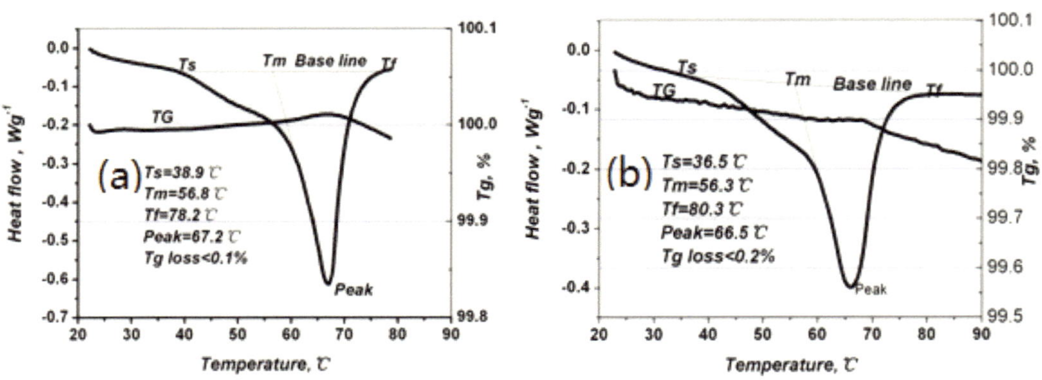

Figure 11. The DSC-TG curves of the paraffin (a) and the PCM/silica composite (b).

56.3°C is 165.16 kJ/kg. TG curve indicates that weight of the composite changes very a few; weight loss of the composite is less than 0.2%.

4. Applications: thermal protection purposes

Numerical and experimental studies are proposed to predict and investigate the thermal absorption characteristics of porous silica infiltrated with PCM for thermal protection applications. The numerical simulation was performed using a volume-averaging technique, and a finite volume modeling (FVM) was used to discretize the heat diffusion equation. The phase change process was modeled using the enthalpy-porosity method.

4.1. Experimental procedure

Three porous silica cylindrical disks (100 mm in diameter and 10 mm thick) with the three different solid-liquid PCMs (two kinds of paraffin and one kind of xylitol) were fabricated as the solid matrices of the silica-PCM composites according to our recent study. The thermophysical data for these samples are given in **Table 3**.

The composites (100 mm in diameter and 10 mm thick) were then introduced into the experimental setup (see **Figure 12** left). The temperature was recorded as it varied with time. The top and bottom walls of the container were insulated by adiabatic materials. For the sake of validation of the numerical model and assumptions, the temperature of the cold face of composite X98# as a function of time was deduced by the numerical simulation.

4.2. Numerical model

A schematic diagram of the composite and two-dimensional grids is given in **Figure 12**. (Sample dimensions: 180 mm in diameter and 120 mm thick). Due to the symmetry and regularity, the samples were formulated with two-dimensional axissymmetric coordinates

	C58#	C64#	X98#
Sample volume (m^3)	3.05E−3	3.05E−3	3.05E−3
Mass before infiltration (kg)	7.20E−1	7.20E−1	7.20E−1
Mass after infiltration (kg)	2.79	2.83	4.16
PCM (paraffin or xylitol) mass rate (%)	74.19	74.56	82.69
Phase change point (K)	324	329	363
Heat storage capacities ($kJ(kg)^{-1}$)	172	165	198
Density (kgm^{-3})	914.75	927.87	1363.93
Thermal conductivity ($Wm^{-1} K^{-1}$)	0.38	0.38	0.46
Specific heat ($kJ(kg)^{-1} K^{-1}$)	2.4	2.4	3.8

Table 3. Properties of the porous silica matrix composites.

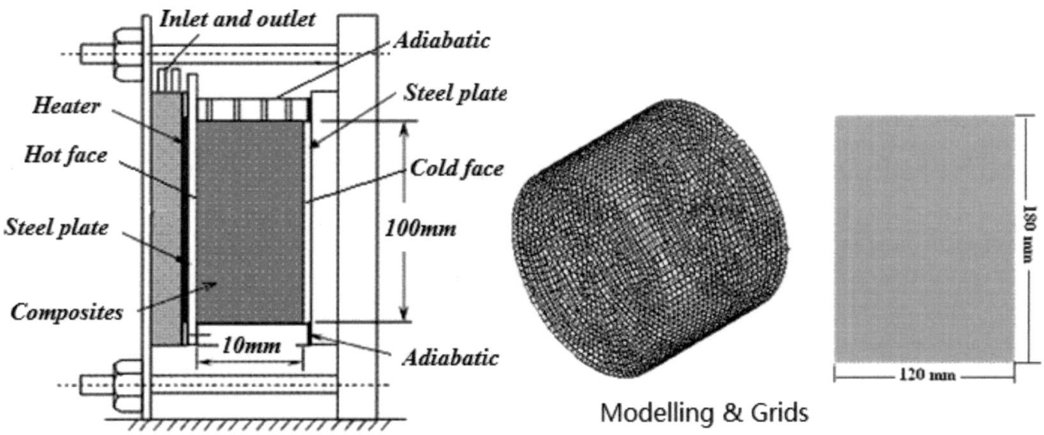

Figure 12. Experimental setup for heat transfer of phase change composite (100 mm in diameter and 10 mm thick) and modeling and grids for the composite (180 mm in diameter and 120 mm thick).

and uniformly split quadrangle grids in all 19,481 nodes by Gambit 2.2.30, and the boundary conditions were specified.

4.2.1. Assumptions and governing equations

The two-dimensional governing equations have been made based on the following assumptions:

The thermophysical properties are different for the solid and liquid phases but are dependent of temperature.

Heat loss from the container to its surroundings is negligibly small.

Based on these assumptions, the enthalpy-porosity method [20] was used for the phase change region in the PCM.

Continuity:

$$\frac{\partial \rho}{\partial t} + \frac{\partial(\rho u)}{\partial x} + \frac{\partial(\rho v)}{\partial y} = 0 \tag{6}$$

where ρ is the density, u is the velocity in u direction, t is the time, and v is the velocity in v direction.

u-Momentum:

$$\rho\left(\frac{\partial u}{\partial t} + u\frac{\partial u}{\partial x} + v\frac{\partial u}{\partial y}\right) = \mu\left(\frac{\partial^2 u}{\partial x^2} + \frac{\partial^2 u}{\partial y^2}\right) - \frac{\partial p}{\partial x} + S_u \tag{7}$$

v–Momentum:

$$\rho\left(\frac{\partial v}{\partial t}+u\frac{\partial v}{\partial x}+v\frac{\partial v}{\partial y}\right)=\mu\left(\frac{\partial^2 v}{\partial x^2}+\frac{\partial^2 v}{\partial y^2}\right)-\frac{\partial p}{\partial y}+S_v \tag{8}$$

Energy:

$$\rho\left(\frac{\partial H}{\partial t}+u\frac{\partial H}{\partial x}+v\frac{\partial H}{\partial y}\right)=\frac{k}{c_p}\left(\frac{\partial^2 H}{\partial x^2}+\frac{\partial^2 H}{\partial y^2}\right)+S_k \tag{9}$$

where the source term is $S_h=\frac{\rho}{c_p}\frac{\partial(\Delta H)}{\partial t}$, and

$$\begin{gathered} H=h+\Delta H \\ h=h_{ref}+\int_{T_{ref}}^{T} c_p dT \end{gathered} \tag{10}$$

where H is the enthalpy, T is the temperature, and ΔH is the specific enthalpy.

The density and dynamic viscosity of the liquid PCM depend on its temperature [20, 21].

Boundary conditions

Given the constant temperature of the cold face and the adiabatic top and bottom wall,

$$\begin{gathered} T(x,y,t)|_{x=d}=T_{cold} \\ \frac{\partial T}{\partial y}|_{y=0,\,t}=0 \end{gathered} \tag{11}$$

Initial conditions

Assuming initial temperature conformity and a fixed temperature of the hot face,

$$\begin{gathered} T(x,y,t)|_{t=0}=T \\ T(x,y,t)|_{x=0,\,t>0}=T_{hot} \end{gathered} \tag{12}$$

The numerical solution was carried out using the Fluent 6.3 software package [19]. The number of computational grids was ~19,200 for the 2D model after grid-independent tests. The time step in the simulations was as small as $\Delta t = 0.1$ s. The convergence was also checked at each time step, with convergence criteria of 10^{-4} for velocity components and 10^{-7} for energy equation.

4.3. Experimental results and discussion

The composites had different thermophysical data owing to the three types of organic PCMs used during the preparation of the porous silica-PCM composites (see **Figure 4** and **Table 3**). The varying heat storage capacities of the composites have an influence on the experimental melting process. Experiments are required to determine the suitable silica-PCM for the service conditions.

The thermal protection properties of the three types of porous silica-PCM were experimentally studied (**Figure 13**). **Figure 13a** compares the temperature of the cold face of the composites when they are subjected to same hot face temperature; **Figure 13b** represents that their temperature of cold face varies with time as they are subjected to hot face temperature (600°C). As can be seen in the figure, the X98# composite shows the best heat absorption properties among the three composites. On the one hand, the temperature of the cold face of the composite X98# is 45°C, while one of the other two composites is 75°C as the three types of composites subjected to same hot face temperature (400°C) (see **Figure 13a**). On the other hand, the slope of the time-temperature curve of the X98# composite decreases at a temperature ~ 80–90°C, which is close to the phase transition range of the X98# composite and does not vary substantially with time. This trend is obvious at the melting temperature of the X98# composite, and it did not appear for the other two composites 58# and C64#, as they are above their phase change temperature and had already melted (see **Figure 13b**). For the cold face of composite X98#, good agreement is obtained when comparing the temperature as a function of time to the numerical simulation results (**Figure 13b**).

4.4. Results and discussion

Numerical studies have been carried out for the three types of composites under identical conditions. The temperature contours for the composite C58# at a constant hot face temperature (573 K), and different time intervals (10, 30, and 60s) are shown in **Figure 14**. The temperature rises quickly at the beginning of the heating process and propagated as time goes on, as the melting of the composites started at the hot face in direct contact with the heating surface and the solid-liquid interface moved gradually in the axial direction over time. For the composite C64#, the heat transfer numerical studies showed the same melting process

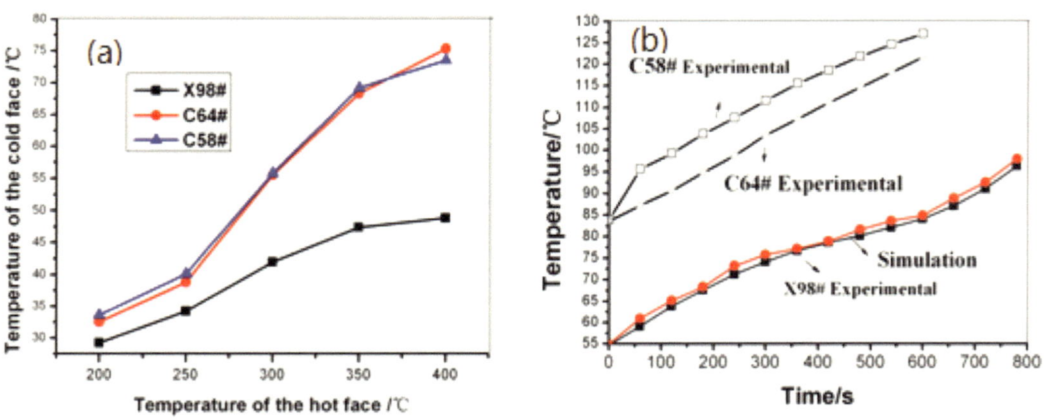

Figure 13. (a) Comparison of cold face and hot face temperature of three types of composites. (b) Cold face temperature varied with time as hot face temperature is 600°C.

compared with the composite C58#. For the composite X98#, the temperature contours at different time intervals (10, 30, 60, and 120 s) are shown in **Figure 15** under identical conditions (a constant hot face temperature (573 K)), and liquid fraction contours of the composites X98# at 180 s are shown in **Figure 16**, which indicated the more stable thermal performance during

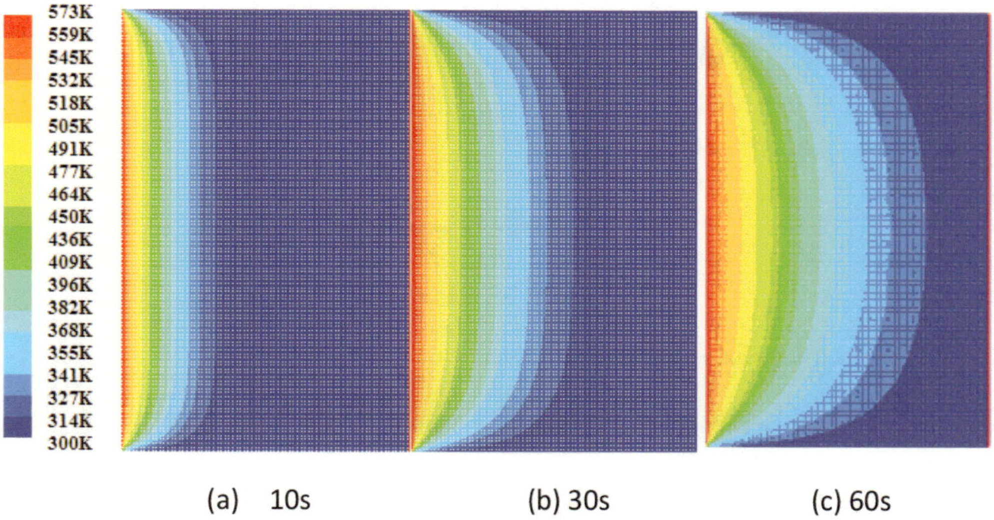

Figure 14. Contours of temperature of the composite C58# (time intervals = (a) 10s, (b) 30s, (c) 60s).

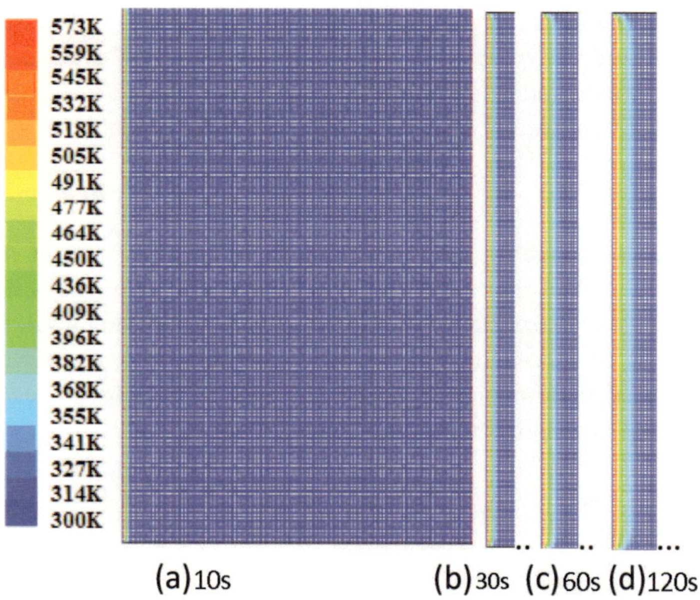

Figure 15. Contours of temperature of the composite 98# (time intervals = (a) 10s, (b) 30s, (c) 60s, (d) 120 s) (left).

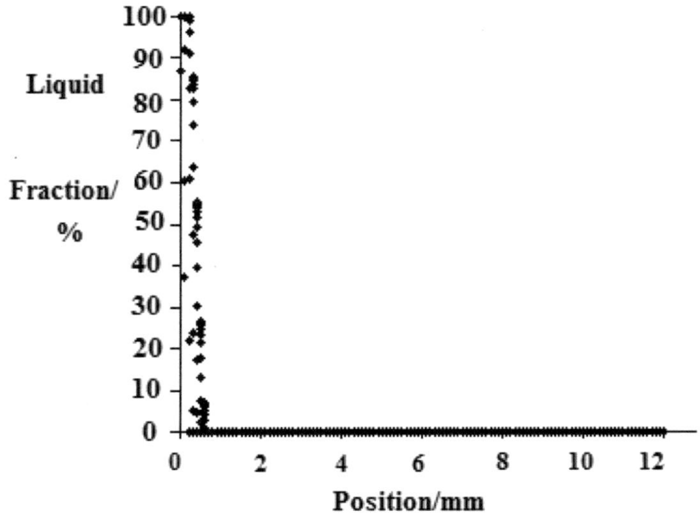

Figure 16. Liquid fraction of the composite 98[#] at 180 s (right).

the melting process due to higher heat storage capacities and phase change point as compared with the composites C58[#] and C64[#].

The experimental results indicated that the suitable silica-PCM for the service conditions is the composite X98[#]. In addition, the composite X98# showed the more stable thermal performance during the melting process based on numerical results for the temperature and liquid contours of the composites. From the thermophysical data of the porous silica matrix composites (**Table 3**), the mass infiltration rate of PCM (xylitol) in the porous silica matrix composite X98[#] is more than the other two composites C58[#] and C64[#]. The X98[#]-type composite has both high heat storage capacity and a high phase change point. Thus, it may be concluded that controlling the surface temperature depends mainly on these two main parameters.

4.5. Conclusions

The effects of the heat storage capacity and thermal properties of porous silica filled with different PCMs were studied numerically and experimentally. The results indicate that the heat storage capacity and phase change point of the composites play important roles in their thermal performance. It has been illustrated that a higher heat storage capacity leads to more stability in the thermal performance of the composite and the phase change point of the composite determined its service conditions.

Acknowledgements

This project is funded by the China Scholarship Council (CSC). The author would like to acknowledge the American Association for the Advancement of Science and Nano Technology and New Materials Institute of Changsha University in China.

Author details

Xiangfa Zhou

Address all correspondence to: flucky-zhou@163.net

Institute of Nano-Technology and New Materials, Changsha University, Hunan, China

References

[1] Dincer I, Rosen MA. Thermal Energy Storage Systems and Applications. England: Wiley; 2002

[2] Su JC, Liu PS. A novel solid-solid phase change heat storage material with polyurethane block copolymer structure. Energy Conversion and Management. 2006;**47**(18–19):3185-3191

[3] Jiang Y, Ding EY, Li GK. Study on transition characteristics of PEG/CDA solid–solid phase change materials. Polymer. 2002;**43**(1):117-120

[4] Grandi S, Magistris A, Mustarelli P, Quartarone E, Tomasi C, Meda L. Synthesis and characterization of SiO_2-PEG hybrid materials. Journal of Non-Crystalline Solids. 2006;**352**(3):273-280

[5] Wang W, Wang CY, Li W, Fan XX, Wu ZH, Zheng J, Li XG. Novel phase change behavior of n-eicosane in nanoporous silica: Emulsion template preparation and structure characterization using small angle X-ray scattering. Physical Chemistry. 2013;**15**(34):14390-14395

[6] Tabata M, Adachi I, Ishii Y, Kawai H, Sumiyoshi T, Yokogawa H. Development of transparent silica aerogel over a wide range of densities. Nuclear Instruments and Methods A. 2010;**623**(1):339-341

[7] Wagh PB, Ingale SV, Gupta SC. Comparison of hydrophobicity studies of silica aerogels using contact angle measurements with water drop method and adsorbed water content measurements made by Karl Fischer's titration method. Journal of Sol-Gel Science and Technology. 2010;**55**(1):73–78

[8] Reynolds JG, Coronado PR, Hrubesh LW. Hydrophobic aerogels for oil-spill cleanup—Intrinsic absorbing properties. Energy Sources. 2001;**23**(9):831–843

[9] Husing N, Schubert U. Aerogels airy materials: Chemistry, structure, and properties. Angewandte Chemie, International Edition. 1998;**37**(1–2):23-45

[10] Zhang ZH, Shen J, Ni XY, Wu GM, Zhou B, Yang MX, Gu XH, Qian MJ, Wu YH. Hydrophobic silica aerogels strengthened with nonwoven fibers. Journal of Macromolecular Science, Part A. 2006;**43**(11):1663-1670

[11] Harel E, Granwehr J, Juliette A. Seeley Alex Pines. Multiphase imaging of gas flow in a nanoporous material remote-detection NMR. Nature Materials. 2006;**5**:321

[12] Morris CA, Anderson ML, Stroud RM, Merzbacher CI, Rolison DR. Silica sol as a nanoglue: Flexible synthesis of composite aerogels. Science. 1999;**284**:622-624

[13] Bi C, Tang GH, Hu ZJ, Yang HL, Li JN. Coupling model for heat transfer between solid and gas phases in aerogel and experimental investigation. International Journal of Heat and Mass Transfer. 2014;**79**:126-136

[14] Kim TY, Hyun BS, Lee JJ and Rhee J. Numerical study of the spacecraft thermal control hardware combining solid-liquid phase change material and a heat pipe. Aerosp Sci Technol, 2013;**27**(1):10-16

[15] Pal D, Joshi YK. Thermal management of an avionics module using solid-liquid phase-change materials. Journal of Thermophysics and Heat Transfer. 1998;**12**(2):256-262

[16] Gregg SG, Sing KSW. Adsorption Surface Area and Porosity. New York: Academic Press; 1982. p. 236-238

[17] Barrett EP, Joyner LG, Halenda P. The determination of pore volume and area distributions in porous substances: Computations from nitrogen isotherms. Journal of the American Chemical Society. 1951;**73**:373-380

[18] Sing KSW, Everett DH, Haul RAW. Reporting physic sorption data for gas/solid systems with special reference to the determination of surface area and porosity. Pure and Applied Chemistry. 1985;**57**:603

[19] Venkateswara Rao A, Sharad DB. Synthesis and physical properties of TEOS-based silica aerogels prepared by two step (acid-base) sol-gel process. Solid State Sciences. 2004;**6**: 945-952

[20] http://www.fluent.com

[21] Reid RC, Prausnitz JM, Poling BE. The Properties of Gases and Liquids. New York: McGraw-Hill; 1987. p. 439-456